The Institute of Biology's
Studies in Biology no. 83

The Genetic Code
and Protein Biosynthesis

Second edition

Brian F. C. Clark

B.A., M.A., Ph.D., Sc.D.

Professor in Biostructural Chemistry,
University of Aarhus, Denmark

Edward Arnold

© Brian F. C. Clark 1984

First published in Great Britain 1977
by Edward Arnold (Publishers) Ltd
41 Bedford Square, London WC1B 3DQ

Edward Arnold (Australia) Pty Ltd
80 Waverley Road,
Caulfield East,
Victoria 3145, Australia

Edward Arnold
300 North Charles Street
Baltimore,
Maryland 21201
U.S.A.

Reprinted 1979
Second edition 1984

British Library Cataloguing in Publication Data

Clark, Brian F.C.
    The genetic code and protein biosynthesis.—2nd ed.
    —(Studies in biology/Institute of Biology; no 83)
    1. Genetic code
    I. Title   II. Series
    574.8'732   QH450.2

ISBN 0-7131-2887-9

To Margaret

Text set in Times 9/11pt
by The Castlefield Press
Printed and bound in Great Britain at
The Camelot Press Ltd, Southampton

General Preface to the Series

Because it is no longer possible for one textbook to cover the whole field of biology while remaining sufficiently up to date, the Institute of Biology proposed this series so that teachers and students can learn about significant developments. The enthusiastic acceptance of 'Studies in Biology' shows that the books are providing authoritative views of biological topics.

The features of the series include the attention given to methods, the selected list of books for further reading and, wherever possible, suggestions for practical work.

Readers' comments will be welcomed by the Institute.

1984
                                                        Institute of Biology
                                                        20 Queensberry Place
                                                        London SW7 2DZ

Preface

When this volume was written in the mid 1970s, we had the general opinion that most of the details of the coding and transfer of genetic information was understood. Although the discovery of the enzyme reverse transcriptase had modified the basic Central Dogma so that it was accepted that RNA could carry the information for the synthesis of DNA, the general complacency in the field considered that the future would be just concerned with filling minor details into the broadly understood pattern. How wrong we were with respect to the eukaryotic systems of higher organisms. The first edition's description of the elucidation of the code and of the basic molecular mechanism of protein biosynthesis is still true for the case of prokaryotic bacteria. That is why the experiments delineating the nature of the genetic code worked. However, as I have described in the appropriate chapters, we have to modify some of our fundamental ideas with respect to eukaryotic systems and must consider as special cases the properties of bacterial and animal viruses concerning colinearity of gene and gene products, the non-overlapping nature of the code and even perhaps the universality of the code. In addition, I have updated where necessary our knowledge on the molecular mechanism of protein biosynthesis. We still believe that the basic molecular details of protein biosynthesis are the same for bacteria and higher organisms.

Denmark, 1984   B.F.C.C.

Contents

1 Cellular Macromolecules

1.1 Opening remarks

The identification of DNA as the genetic material and the establishment of its double helical structure can be considered as starting points in the growth of the new discipline of molecular biology. The structure proposed by Watson and Crick readily indicated how DNA could be duplicated to ensure hereditary continuity during cell division. However, a decade elapsed before it could be established how DNA could also control the cell's metabolic processes by providing information for the synthesis of cellular enzymes. When it was realized in the early 1950s that proteins were gene products and not genetic material, a search began for the link between the two polymer languages – the nucleic acid and protein languages. This relationship, the genetic code, is in a sense the key to molecular biology. Its elucidation is one of the major achievements of twentieth-century science and was carried out in a highly competitive atmosphere at several research centres (Table 1–1). Before I describe in detail the solving of the code and more recent knowledge about signals used in its expression, it will be useful to describe the protein and nucleic acid structures and the place they have in biological systems.

1.2 Levels of molecular organization

The basic unit of biological material is the cell. It can be free living, as for example the bacterium *Escherichia coli*, or in a special organization with neighbours, as in mammalian tissues and organs. What gives the cell its property of life is not known. However, all the scientific evidence available indicates that a cell is made up from inanimate molecules. Nearly 30% of the cell's mass consists of macromolecules which contain a series of subunits in their structures. The macromolecules in the cell function as structural parts, energy stores, repositories of genetic information and as special molecules for controlling the chemical processes keeping the cell alive. Chemists and biochemists know much less about the physics and chemistry of macromolecules than about their components, but we are learning rapidly as the difficulties of handling large molecules are overcome. Indeed, we know enough about protein and nucleic acid structures to talk sensibly in chemical terms about their relationship in the elucidation of the genetic code. Two generalizations about structural concepts can be made at this stage in our knowledge. First, the physical and chemical properties of a macromolecule are not just a summation of the subunits' properties; these can change because of neighbouring group and spatial interactions. Secondly, as far as we know, biological macromolecules and molecules obey the laws of physics and

Table 1–1 Important events in the elucidation of the genetic code.

Discovery	Date	Research workers	Place of work
DNA transforms bacteria	1944	O. T. Avery, C. M. MacLeod & M. McCarty	Rockefeller Institute, New York
DNA is viral genetic material	1952	A. D. Hershey & M. Chase	Cold Spring Harbor Laboratory, U.S.A.
Double helical model for DNA	1953	J. D. Watson & F. H. C. Crick	Cavendish Laboratory, Cambridge University
		M. Wilkins	King's College, London
Primary structure of a protein	1954–55	F. Sanger	Biochemistry Department, Cambridge University
Adaptor hypothesis	1957	F. H. C. Crick	Cavendish Laboratory, Cambridge University
Discovery of tRNA as the adaptor	1958	M. B. Hoagland	Massachusetts General Hospital, Boston
Ribosomes as templates for protein synthesis	1958–59	J. D. Watson & A. Tissieres	Harvard University, U.S.A.
Postulation of messenger RNA	1960–61	F. Jacob & J. Monod	Institut Pasteur, Paris
		F. Gros	Harvard University, U.S.A.
		S. Brenner & M. Meselson	California Institute of Technology, U.S.A.
Cell-free system for translation	1961	M. W. Nirenberg & H. Matthaei	National Institutes of Health, Bethesda, Maryland, U.S.A.
		S. Ochoa & colleagues	New York University Medical School
Properties of the code from bacterial and bacteriophage genetics	1961–62	F. H. C. Crick, S. Brenner & colleagues	MRC Laboratory of Molecular Biology, Cambridge
Direction of synthesis of proteins	1961–62	H. Dintzis	MIT, Cambridge, U.S.A.

Discovery	Date	Research workers	Place of work
Polysomes	1963	A. Rich & colleagues	MIT, Cambridge, U.S.A.
Colinearity of gene and protein	1964	A. Gierer	Max Planck Institute, Tübingen, Germany
		C. Yanofsky & colleagues	Stanford University, California, U.S.A.
		S. Brenner & colleagues	MRC Laboratory of Molecular Biology, Cambridge
Triplet binding assay	1964	M. W. Nirenberg & P. Leder	National Institutes of Health, Bethesda, Maryland, U.S.A.
A special tRNA for initiation	1964	K. A. Marcker & F. Sanger	MRC Laboratory of Molecular Biology, Cambridge
Translation of mRNA of defined sequence	1965	H. G. Khorana & colleagues	University of Wisconsin, Madison, U.S.A.
Primary structure of a tRNA	1965	R. W. Holley & colleagues	Cornell University, U.S.A.
		H. G. Zachau & colleagues	Institute of Genetics, University of Cologne, Germany
Initiation codons	1965–66	B. F. C. Clark & K. A. Marcker	MRC Laboratory of Molecular Biology, Cambridge
Termination codons	1965–66	S. Brenner & co-workers	MRC Laboratory of Molecular Biology, Cambridge
		A. Garen & co-workers	Yale University, U.S.A.
Wobble hypothesis	1966	F. H. C. Crick	MRC Laboratory of Molecular Biology, Cambridge
Tertiary structure of tRNA	1974	A. Klug, B. F. C. Clark & colleagues	MRC Laboratory of Molecular Biology, Cambridge
		A. Rich, S. Kim & colleagues	MIT, Cambridge, U.S.A.
Overlapping codon usage	1976	B. G. Barrel, F. Sanger & colleagues	MRC Laboratory of Molecular Biology, Cambridge
Introns and exons	1977	R. Flavell, P. Chambon, P. Leder & colleagues	Amsterdam, Strasbourg and NIH, Bethesda, U.S.A.
Z-DNA	1979	A. Rich & colleagues	MIT, Cambridge, U.S.A.

chemistry so we do not need new concepts to investigate their structure and function.

The generally accepted levels of molecular organization constituting a cell are shown diagrammatically in Fig. 1–1. Simple compounds such as carbon dioxide and water are taken from the environment and converted *via* small molecular weight (50–250)* intermediates into macromolecular building blocks. Different subunits of the order of 100–350 in molecular weight are polymerized into their respective macromolecules as shown. The simple building blocks, or monomers, are linked covalently with chemical bonds into their respective polymeric macromolecules. The macromolecules range widely in molecular weight from 10^3 to 10^9. The range for the lipids is much lower (up to 2500) than for the other macromolecules, but they are usually considered as macromolecules because they play a role in macromolecular assemblies such as

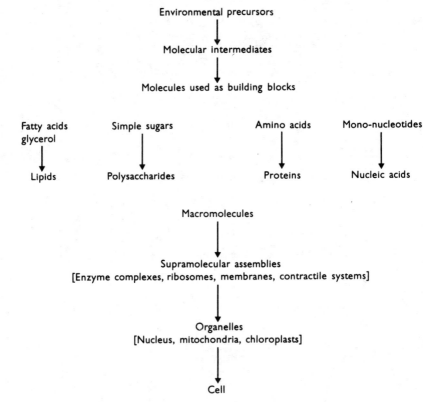

Fig. 1–1   Levels of molecular organization.

* Molecular weights are always given in terms of daltons, either abbreviated to d or omitted altogether, as above. One dalton is the mass of one hydrogen atom, i.e. 1.67×10^{-24} g.

membranes. There is an important difference between nucleic acids and proteins on the one hand and carbohydrates and lipids on the other. Nucleic acids and proteins are considered to be informational macromolecules because of their structure. Most nucleic acids contain four different types of subunit and proteins twenty. The arrangement of these subunits in linear sequence contains genetic information, but whereas the nucleic acids can be the genetic material or genetic intermediaries, the proteins are the gene products. Indeed, a gene is now normally defined as that piece of a nucleic acid carrying a unit of genetic information, usually for a protein but also for other nucleic acids. In contrast to the informational macromolecules, the non-informational polysaccharides and lipids contain repeats of identical subunits.

The four major types of macromolecule have identical functions in all cell species. The nucleic acids store and transmit genetic information. The proteins are the direct products and effectors of gene action and incorporate genetic information. Most proteins have a specific catalytic activity as functional enzymes but others serve as structural elements. Proteins are the most flexibly versatile of the macromolecules so they can have many other biological functions. Both polysaccharides and lipids can serve in the cell as fuel storage elements and as structural components.

Macromolecules organize themselves into supramolecular assemblies apparently without new genetic information or extra influences. These supramolecular assemblies, such as, for example, ribosomes which are nucleoproteins, can be so large that they are insoluble particles in the cell. Complexes of proteins and lipids also take part in organized structures as membranes; their molecular weights can range from 10^3 to 10^9. An important aspect of supramolecular assemblies is that they are not attached by covalent bonds but are held together by a combination of electrostatic and numerous weak bonds, weakly electrostatic hydrogen (H)-bonds, or non-polar (hydrophobic) interactions together with special stereochemical fits like 'hand-in-glove' spatial relationships. This means that they can be dissociated into components fairly easily. It is likely that these types of non-covalent bonding interactions form the basis of the macromolecular architecture and interactions in the cell.

1.3   Cellular components

The molecular make-up of cells does not vary radically from type to type in general. Thus the bacterial cell components shown in Table 1–2 reflect average components of most cells. Of course there are exceptions, especially cells with very specialized function and cells which form non-living parts of organisms, such as fat storage material or bone, hair and feathers. Apart from water, which keeps most of the cellular components in solution, macromolecules are the most abundant of the cellular components, and of them proteins as a class constitute the largest part. Proteins have a tremendous range of function and molecular weight so that the figure of 40 000 to 100 000 for the average molecular weight in Table 1–2 refers to the majority of proteins. The nucleic acids are subclassified according to structure and function as shown by their abbreviations in Table 1–2. More explanation will be found in section 1.5. Other cell components such as carbohydrates and lipids are not concerned with the transmission of genetic information.

Table 1–2 Cellular components of a typical bacterial cell.

Molecular component	Subclassifications of components		% total weight	Mol. wt. average	No. of different kinds
Proteins			15	$4\text{–}10 \times 10^4$	3000
Nucleic acids	DNA		1	2.5×10^9	1
	RNA		6		
		mRNA		10^6	1000
		tRNA		2.5×10^4	~ 50
	23S	rRNA		1.1×10^6	1
	16S	rRNA		5.5×10^5	1
	5S	rRNA		4×10^4	1
Carbohydrates			3		~ 50
Lipids			2		~ 40
Small organic molecules			2		~ 500
Inorganic ions			1		10–20
Water			70	18	1

1.4 Proteins

Proteins form a major constituent of living organisms. Indeed, their complexity of function and structure is such that they can be considered to be an attribute of life itself. They have five main functions: as enzymes, antibodies, structural elements, transport devices and metabolic regulators. As *enzymes* they have catalytic properties controlling the building up and breaking down of organic molecules (metabolism) and in releasing energy from incoming molecules to provide power for essential life processes. *Antibodies* are large complicated assemblies of protein subunits concerned with the defence of the organism against foreign bodies. The protein *structural elements* define and maintain the architecture of the cell. It is also noteworthy that proteins form structural capsules for the genetic material in simple bacterial viruses called bacteriophage, which are considered to be on the borderline of life. Bacteriophage cannot be called living unless they infect a bacterial cell where they can then reproduce. As *transport devices* proteins carry small molecules or ions within the cell or through membranes. Finally, as *metabolic regulators* proteins can co-ordinate and direct the chemical processes in the cell, acting either indirectly on genetic information as hormones or directly as repressor molecules. Apparently proteins are concerned with all biological processes within the cell. Since the functions are so diverse, the chemistry is expected to be very complicated but knowledge of the structure of proteins has progressed far enough to explain much of this diversity. Proteins do not contain genetic information in a form for transmission to other systems; they are the end products of genetic information. Many of their functions are short-lived and the molecules themselves can be synthesized rapidly and degraded within the cell – not a desirable property for carriers of genetic information because too many mistakes could arise. Thus the role of genetic material is given to the very stable molecules of DNA. Its structure will be discussed in section 1.5. The genetic information stored in DNA needs to be expressed in protein function; the transmission and deciphering of genetic information is described in Chapter 2.

Structure

The great range of functional diversity of proteins is reflected in their structural diversity. However, two structural features are common to all proteins: they contain varying proportions of only 20 different types of amino acids, the fundamental, common or standard amino acids; and they are polypeptides, polymers of amino acids joined in a regular fashion by peptide bonds. Previously, molecular biologists sometimes used the term essential instead of common for the 20 amino acids. Unfortunately, this conflicts with the biochemist's idea of essential amino acids, where these amino acids cannot be synthesized by, but need to be supplied to, an organism.

Although many proteins contain 300 to 400 amino acid units, they may range in size from about 50 units in the case of insulin to 2500 units in the case of the muscle protein myosin, and to even larger complexes of molecular weights of several millions. When proteins are degraded by acidic hydrolysis or by proteolytic enzymes, the common amino acids are released although they need not all be present. Occasionally some non-standard amino acids are found in proteins but these are derivatives of the common 20, formed after the protein has been biosynthesized from the common 20.

The structural formulae and three-letter abbreviations usually used for the amino acids are given in Fig. 1–2. The common 20 amino acids are α-amino acids with one exception, proline, which is an α-*imino* acid. It has an –NH– (imino) group in the α position next to the –COOH (carboxyl) group instead of an –NH$_2$ (amino) group. The general structural formula RCH (NH$_2$) COOH for the α-amino acids shows that they have a common structural part and differ only in the R group. The different R groups are shown in Fig. 1–2 and are divided into three general classes. They give variety in properties to the amino acids and hence to the protein structures that the amino acids constitute.

The first class of *non-polar* or *hydrophobic* amino acids includes those with aliphatic side chains (alanine, leucine, isoleucine, valine and proline), those with aromatic side chains (phenylalanine and tryptophan) and one containing sulphur (methionine). They are characterized by not having an affinity for water and tend to put their R groups together to exclude water. This is an important property for the folding of proteins and for making lipoprotein structures. The second class of *uncharged polar* amino acids includes the hydroxyl (OH)-containing serine, threonine and tyrosine; amide-containing asparagine and glutamine and sulphur-containing cysteine. It also includes, for convenience, glycine which is the smallest amino acid with no sharply distinctive properties and in which R is an atom of hydrogen. This class has an affinity for water which tends to make the protein soluble. The third group of *charged polar* amino acids includes amino acids which are negatively charged at the physiological pH of around 7, aspartic and glutamic acids, and those which are positively charged, arginine, lysine and histidine, albeit the last very weakly.

One general property of the amino acids, which will only be mentioned briefly, is optical activity. Biological molecules are usually asymmetric and have optical activity, i.e. they can rotate the plane of polarized light. In the case of amino acids, all of them except glycine have four different chemical groups attached to the α-carbon, making them asymmetric molecules, so that they are optically active. Fortunately for the research chemist, all of these 19 found in nature have the same kind of asymmetry so that their opposites (enantiomorphs) do not occur in proteins although some do occur in other

$$R$$
General structure is NH_2—$\underset{\alpha}{CH}$—$COOH$ The α carbon atom is indicated

Class	Name	R-group	Abbreviation
Non polar or hydrophobic	Alanine	CH_3—	Ala
	Leucine	CH_3—CH—CH_2— with CH_3	Leu
Aliphatic	Isoleucine	CH_3CH_2—CH— with CH_3	Ile
	Valine	CH_3—CH— with CH_3	Val
	Proline	CH_2—CH_2—to α—N / CH_2—to α—C	Pro
Aromatic	Phenylalanine	⬡—CH_2—	Phe
	Tryptophan	CH_2— (indole ring with NH)	Trp
S-containing	Methionine	CH_3—S—CH_2—CH_2—	Met

		Structure	Abbr.
Uncharged polar	Glycine	H—	Gly
OH-containing	Serine	OH—CH$_2$—	Ser
	Threonine	CH$_3$—CH— (OH)	Thr
	Tyrosine	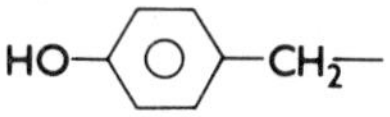 HO—⟨○⟩—CH$_2$—	Tyr
Amide-containing	Asparagine	CONH$_2$—CH$_2$—	Asn
	Glutamine	CONH$_2$—CH$_2$—CH$_2$—	Gln
S-containing	Cysteine	SH—CH$_2$—	Cys
Charged polar Negatively charged	Aspartic acid	COO$^-$—CH$_2$—	Asp
	Glutamic acid	COO$^-$—CH$_2$—CH$_2$—	Glu
Positively charged	Histidine	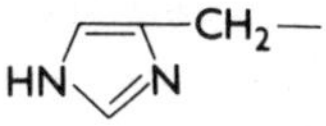 HN⟨═⟩N—CH$_2$—	His
	Arginine	NH_3^+—C(═NH)—NH—CH$_2$—CH$_2$—CH$_2$—	Arg
	Lysine	NH_3^+—CH$_2$—CH$_2$—CH$_2$—CH$_2$—	Lys

Fig. 1–2 The structural formulae and abbreviations of the 20 common amino acids.

molecules. This is also true of the nucleic acid building blocks (see section 1.5). All of the standard common amino acids have turned out to be L-amino acids so we shall omit the L-prefix in our discussions.

All proteins are polypeptides. Thus they are linear arrangements of amino acids joined through peptide bonds as shown in Fig. 1–3, where R$_1$ and R$_2$ signify the R groups belonging to different amino acids. This linear arrangement or sequence of amino acids is termed the primary structure. The two features of the primary structure of proteins relevant to our understanding of the genetic code are (1) the common peptide bonds linking the different amino acids in such a linear or one-dimensional sequence giving a polypeptide *backbone* to the protein and (2) the diversity of structure depending on the different R̊ groups which stick out from the backbone structure. It is the order of the R groups which is genetically determined because then the protein's structure and function is, as far as we know, completely determined. The properties of the different R groups cause the protein to fold up automatically in a three-dimensional structure like that determined for myoglobin in Fig. 1–4 and specify the structure and functional activity of the protein. The huge

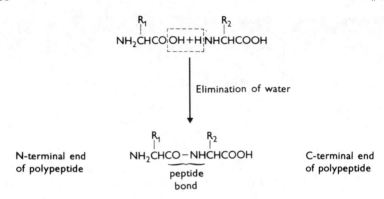

Fig. 1–3 Formation of the peptide bond.

diversity of possible protein structures can be understood when, even for a short polypeptide four amino acids long containing arrangements of 20 different amino acids, $20 \times 20 \times 20 \times 20$ or 160 000 different sequences are possible. For a protein of length 300 we have 20^{300} possible structures, many magnitudes greater than the total number of proteins known. Evolution has plenty of possibilities to work with to attain the catalytic activities of the presently known enzymes.

Our understanding of protein biosynthesis which reflects the deciphering of the genetic code was enormously simplified by the discovery that we had to solve only the relationship between macromolecules containing one-dimensional information. It was thus very likely that, because we need to specify only the primary structure of the protein, a one-dimensional code would be used in the genetic material. Also, because we need to specify only the primary structure of proteins by genetic information, higher levels of structure are outside the scope of this book.

Fred Sanger earned a Nobel prize for the first successful determination of the primary structure of a protein, that of insulin (Table 1–1). Although insulin is relatively small, only 51 amino acids arranged in two chains, its structural determination was immensely important as laying the foundations of protein structure. Before then we did not know that proteins consisted of linear arrays with regular head to tail amino acid junctions. This feature gives rise to the concept of polarity in the primary structure. For example there is usually a free amino group called the *N-terminal end* at one end of the protein and a free carboxyl group at the other, called the *C-terminal end* (Fig. 1–3). Furthermore, it is known from some significant biochemical experiments in the early 1960s that proteins are biosynthesized in cells from the N-terminal end towards the C-terminal end. These experiments involved growing blood cells (reticulocytes) on a culture of the essential amino acids with one of them labelled with radioactive carbon (^{14}C). When a particular protein was isolated from the cells after increasing times of labelling, it was found on analysis that the radioactivity tended to accumulate near the C-terminal end. Since most of

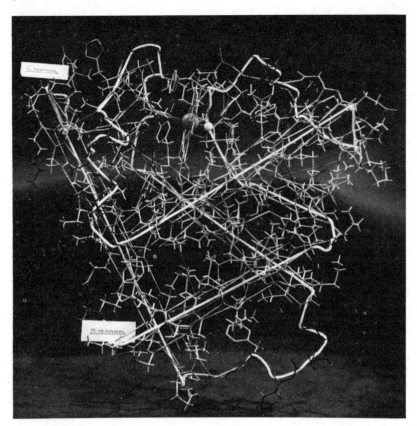

Fig. 1–4 Molecular model of sperm whale myoglobin shown in skeletal form. (Courtesy of Sir John Kendrew, MRC Laboratory of Molecular Biology, Cambridge.)

the new synthesis of the protein had been concerned with the completion of the protein, so that the protein was not being synthesized afresh from the beginning, this result indicated that the C-terminal end is the last part of the polypeptide chain to be made.

To help clarify terminology other levels of protein structure will be briefly mentioned. The distinctions in structural properties are usually made in terms of the interactions necessary for their maintenance. The *primary structure*, as we have seen, is the sequence of amino acids linked together by covalent bonds. *Secondary structure* arises by the ordering of the polypeptide backbone due to short range weakly electrostatic hydrogen (H)-bonds between carboxyl oxygen atoms and amide nitrogen atoms shown by dots in the scheme $> C = O$ .... $H-N <$ where the dashes are covalent bonds. In proteins two classes of structures result: *α-helical* or *β-pleated sheet* structures. In addition it is found that the polypeptide backbone, even if in an α-helix, can be folded into a variety of conformations giving an overall sometimes globular structure which is the *tertiary* or three-dimensional structure. Interactions which hold together the tertiary structure can be H-bonding, salt linkages, disulphide bonds, and weak attractive forces called van der Waals forces. Finally we can have aggregates of individual protein subunits defining a *quaternary* structure for a specific function. Aggregate formation, which in haemoglobin for example is from four subunits, is brought about by surface interactions between R side chains and between exposed backbones resulting in a geometrical fit between subunits. The first proteins to have their three-dimensional structure determined by X-ray crystallographic analysis were myoglobin (an example of tertiary structure of molecular weight 17 000 and shown in Fig. 1–4), and haemoglobin (an example of tertiary and quaternary structure of molecular weight 65 000). These structures were determined by research groups led by J. C. Kendrew and M. F. Perutz (Table 1–1), for which work they shared a Nobel prize.

1.5   Nucleic acids

The nucleic acids, like proteins, are informational macromolecules. However, whereas proteins are gene products the nucleic acids contain genetic information which is duplicated and separately packaged into daughter cells to ensure inherited characteristics and which regulates the synthesis of, or is transformed into, protein structures involved in cellular metabolism. Two main classes of nucleic acids are defined by their constituent sugar (Fig. 1–5); DNA (deoxyribonucleic acid) contains D2'-deoxyribose and RNA (ribonucleic acid) contains D-ribose. D defines the stereochemistry of the ribose shown and is often omitted for convenience. DNA is chemically much more stable than RNA and is thus well suited to its genetic role as a stable physical thread of hereditary continuity. It is usually found packaged in the nucleus of eukaryotes in the form of chromosomes and in a less well defined region, sometimes called a nucleoid, in prokaryotes. RNA acts as genetic material only in special cases, for example in some animal viruses and bacteriophages. RNA is usually involved in the transfer of genetic information to the cellular apparatus where this information is transformed into functional proteins. Three classes of RNA are normally distinguished according to function: messenger RNA (mRNA), ribosomal RNA (rRNA) and transfer RNA (tRNA). Another kind of RNA, regulatory RNA, may exist in the cell for certain signalling events during gene expression but this is not well defined. The types of RNA are generally assumed to be functionally localized in the cytoplasm. Of course they all begin their existence by being copied (transcribed) off the genetic material. Indeed, the genes for eukaryotic rRNA are found in a special region of the nucleus called the nucleolus. Furthermore a new type of very large RNA containing diverse structures and of rather ill-defined function has been shown to exist in the nucleus. This heterogeneous nuclear RNA (hnRNA) may have functions other than strictly precursor messenger RNA (mRNA). There has been recently established evidence that RNAs might also act as enzymes under special circumstances.

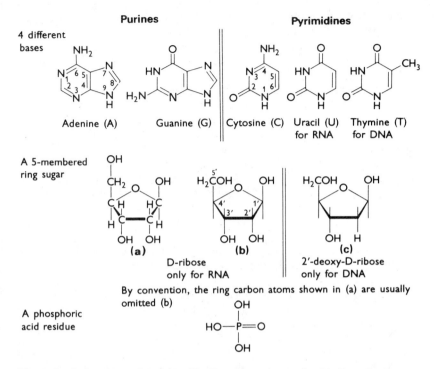

Fig. 1–5 Components of nucleic acids. Base + sugar = nucleoside; base + sugar + phosphoric acid = nucleotide. The numbering systems are shown so that the linkage points in a polynucleotide as in Fig. 1–6 can be readily understood. The bases or usually the total nucleosides are referred to by one letter abbreviations taken from the initial letters of the base names (see Fig. 1–6).

A comparison of the sizes and abundance of DNA and RNA in a typical bacterial cell is given in Table 1–2. While about 50% of the cell's dry weight is protein, about 23% is nucleic acids. There are three separate rRNA species which play a structural role in prokaryotic ribosomes. They are referred to by their sedimentation properties in terms of Svedberg or S units (Table 1–2).

Component structures

Nucleic acids are polymers of nucleotides. Each nucleotide contains a nucleoside attached to a phosphoric acid residue. A nucleoside consists of an aromatic nitrogenous base linked to a pentose sugar. Enzymes which degrade nucleic acids are called nucleases.

Only four different heterocyclic (i.e. containing more than one type of atom) ring bases are found in general in nucleic acids. They fall into two types: the double ring containing type (purines) which includes adenine and guanine and the single ring containing class (pyrimidines) which includes thymine and cytosine in the case of DNA but uracil and cytosine in the case of RNA (Fig. 1–5). The conventional numbering schemes for the bases are given in Fig. 1–5.

In some DNA molecules other bases which are derivatives of the standard four are occasionally found but these are special cases. Furthermore, some classes of RNA such as tRNA and rRNA contain several different types of bases called rare or modified minor bases; these also are usually considered to be derivatives of the standard four.

In spite of the difference in one base constituent the main difference between DNA and RNA is in the sugar constituents whose structures are shown in Fig. 1–5. Nucleosides are formed by joining the bases to the sugars. A ribonucleoside or a deoxyribonucleoside is formed by joining ribose or deoxyribose to a base. For example, adenine + ribose gives adenosine (A) whereas adenine + deoxyribose gives deoxyadenosine (dA). Similarly guanine, cytosine, uracil and thymine give rise to guanosine, cytidine, uridine and thymidine. The C–1′ position of the sugar is joined by an N-glycosyl bond to the N–9 of the purine or N–1 of the pyrimidine to form nucleosides as in Fig. 1–6. In some biochemical texts this type of bond is wrongly called an N-glycosidic bond.

The phosphoric acid residue is joined in a phosphate ester link to the nucleoside to give a nucleotide or nucleoside phosphomonoester of general formula $[ROPO_3H_2]$.

Both DNA and RNA are linear polymers of nucleotides. The primary structure of a piece of RNA is shown in Fig. 1–6. It is equivalent to joining nucleotides with the elimination of water in a series of phosphodiester groups. In a similar piece of DNA the C–2′ OH-group would be replaced by C–2′ H and the uracil by thymine. Shorter ways of writing nucleotide structures in nucleotide sequences are also given in the figure. For example, the sequence of nucleotides shown is usually simply written as pGpApUpC. A nucleic acid's order of sequence of nucleotides is its *primary structure* and has a constant backbone feature as do proteins. Nucleic acids may also have secondary and tertiary structures as we shall see when discussing DNA (section 1.6) and tRNA. With nucleic acids the backbone is a polyphosphodiester and not a polypeptide, with the R groups of amino acids now replaced by bases. The phosphate groups join C–3′ hydroxyl groups to C–5′ hydroxyl groups in all RNAs and DNAs so far identified. These linkage points were determined by a combination of chemical and specific enzymic cleavage methods. The joining of the nucleotides in this regular, defined manner gives polarity to the nucleic acid chain so that the backbone will look different according to the direction along which one looks. Thus there is a standard convention for writing nucleic acid primary structures with the same polarity so that they can be easily compared. The sequence is written from left to right as in pGpApUpC so that the phosphate group, p, joins the C–3′ hydroxyl group of the left side nucleoside to the C–5′ hydroxyl group of the right side nucleoside. As a result the end of the nucleic acid at the left side is called the 5′-end and the right side the 3′-end. This convention will be used throughout the rest of this book.

Although a nucleic acid backbone specifically joining only sugar hydroxyl groups precludes any branching in a DNA structure, the extra C–2′–hydroxyl group in RNA could give rise to phosphate linkages and thus branched

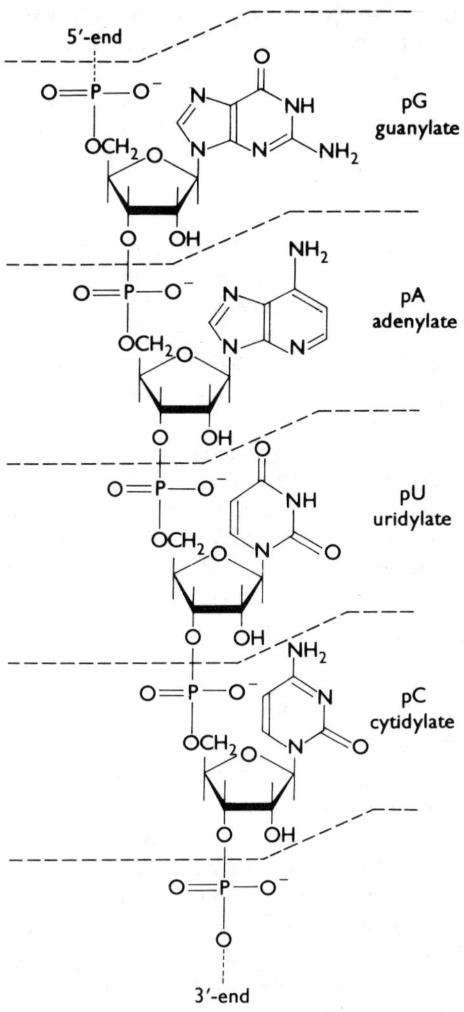

Fig. 1–6   The chemical structure of a piece of RNA.

structures. However, all DNA and RNA molecules so far characterized have linear polymeric structures. Nowadays the most convenient way of determining such linearity is by visualization in an electron microscope.

1.6   The genetic material

The genetic information of eukaryotic cells is carried by nuclear structures called *chromosomes*. These are composed largely of DNA (30% dry weight) and protein. At the time of cell division they take on a dense character that facilitates staining and so allows viewing in the light microscope. Thus for several decades the presence of DNA in cells has been detected by a specific test involving staining (Feulgen reaction) for visualization in the light microscope. Such tests indicated that DNA is a component of chromosomes which are known to be carriers of hereditary information. Evidence suggesting that DNA is the genetic material comes from observations that the content of DNA in a cell varies with the ploidy (the multiplicity of chromosomal information). For example, haploid cells have half the DNA content of diploid cells. Much molecular detail is now known about the organization of the DNA and protein in a chromosome, but its complete determination is the subject of intense research. So far the disposition of DNA and histone protein components have been determined for the nucleosome, the elementary repeating structural unit of chromatin. A nucleosome contains a piece of DNA of about 160 base pairs length, wrapped about a protein core in helical fashion. The protein core consists of 2 molecules each of 4 different histones. The replication of DNA in such complicated structures remains to be explained.

The way in which the genetic material is replicated in prokaryotic cells is much better understood, since the DNA is not encumbered by structural proteins and there is also lack of apparent multiplicity of DNA fibres. Indeed, the first really convincing evidence that DNA is the genetic material came from experiments in the early 1940s on bacterial transformation which showed that DNA isolated from one species of pneumococcus could enter a second strain which lacked a capsular coat and could convert the second strain into a coat-producing strain.

Secondary structure of DNA

DNA is a linear polydeoxyribonucleotide structure and this one-dimensional primary structure is all we need to consider for the transmission of genetic information. However, many of the nucleic acid structural interactions used in deciphering the genetic code are explained as a result of the knowledge of the secondary structure of DNA. The correct determination of the DNA structure embodied in a model like that shown in Fig. 1–7 earned a Nobel prize for J. D. Watson and F. H. C. Crick, which they shared with M. H. F. Wilkins who provided the crucial experimental evidence. Two types of model are illustrated. The left-hand one is called a space-filling model and gives an idea of the bulk of the atoms. It can readily be seen that DNA is a grooved rod-like structure with no internal empty spaces which could allow penetration by water or small molecules. The right-hand model is shown in a skeletal form so that the chemical bonds (full lines) and hydrogen bonds (dotted lines) can be seen and more easily discussed, but the real structure is more like the left-hand picture. Biological macromolecules function in respect of their spatial structures. However, the spatial structure of DNA, which is important in terms of its function as genetic material, is defined by secondary structure – an arrangement of the linear structure held together by short distance weak forces called hydrogen bonds and by affinity between overlapping neighbouring bases called stacking forces.

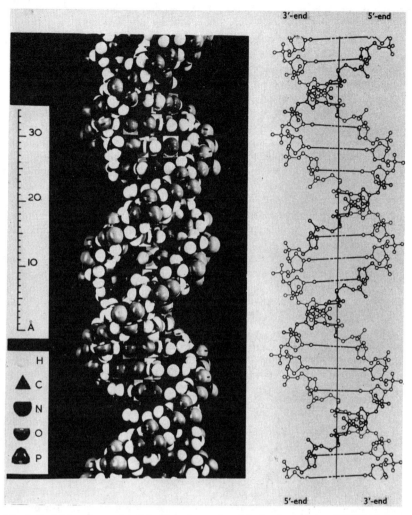

Fig. 1–7 Molecular model of the B conformation of deoxyribonucleic acid (DNA). See text for details. (Courtesy of Professor M. H. F. Wilkins, King's College, London.)

The model for DNA secondary structure arrived at by Watson and Crick in 1953 made sense of a number of experimental observations which can only be briefly mentioned here. The structural components shown in Fig. 1–5 were known. Physical chemical experiments such as measurements of viscosity and molecular weight and titration with acid or base suggested that DNA is a rod-like molecule but of more than one subunit held together by H-bonding. Furthermore, Chargaff determined that there is a base equivalence in the structure. After determining the base compositions of many DNA samples from different biological materials, he found that the amount of purine base expressed in molar quantities was always the same as that of pyrimidine base. Moreover, the amount of base A always equalled base T and that of base G equalled C. The final crucial evidence came from X-ray diffraction studies by Wilkins and his colleagues working with fibre preparations of DNA. This evidence of the order of the structure gave quantitative details of the disposition of the bases.

The resulting model shown in Fig. 1–7 describes a DNA molecule in complete structural detail; its accuracy has stood the test of time, but only in 1981 has its molecular details been proved by X-ray crystallographic studies of short pieces (12 subunits long) of synthetic DNA. First it is necessary to understand that a DNA molecule consists of not just one but two polydeoxyribonucleotide chains winding about each other in a double stranded right handed helix. The DNA molecule thus defined is a double helical rod-like structure. The polarities of the two chains (also called strands) are antiparallel. Thus if the model is viewed from the bottom upwards (along one axis), the phosphodiester bonds go in opposite directions on opposite strands (e.g. in Fig. 1–7 the bonds on the left-hand chains at the bottom of the right-hand model go $3' \rightarrow 5'$ in one chain and $5' \rightarrow 3'$ in the other). The model is constructed so that the phosphate-deoxyribose backbone making the helical conformation winds about an axis with a constant radius of 1 nm. The constant distance apart of the two backbones is achieved by arranging bases protruding from each backbone perpendicular to the axis so that a purine is always opposite a pyrimidine. Furthermore, to correlate the structure with the experimental observations, A is placed opposite T and G opposite C. In this configuration the predicted hydrogen bonding is seen to link these two sets of bases, now called base pairs because of their complementary nature in the structural determination. Because they are found in DNA these base pairs illustrated in Fig. 1–8 are also called standard or Watson-Crick base pairs. The distances involved in the hydrogen bonded base pairs are shown in the figure. The constant distance between the sugars joining to the bases is also shown. It can be seen that a G and C base pair is more strongly held together than an A:T because three hydrogen bonds rather than two are involved. An X-ray identified repeat of 0.34 nm corresponds to the distance between successive bases in one chain, i.e. distance between base pairs along the axis. The pitch of the helix was found to be 3.4 nm giving 10 base pairs per complete turn of the helix. Since the DNA molecule is very long, the number of hydrogen bonds holding the two strands together is very great, making a very stable double stranded molecule. For some time it was thought that the hydrogen bonds determined the helical conformation, i.e. the secondary structure. However, vaguely defined hydrophobic interactions caused by aromatic electronic orbital overlapping between successive bases probably contribute more to the helical conformation because it is possible to stabilize single stranded polynucleotide helices. These interactions between successive bases, as noted previously, are called stacking interactions.

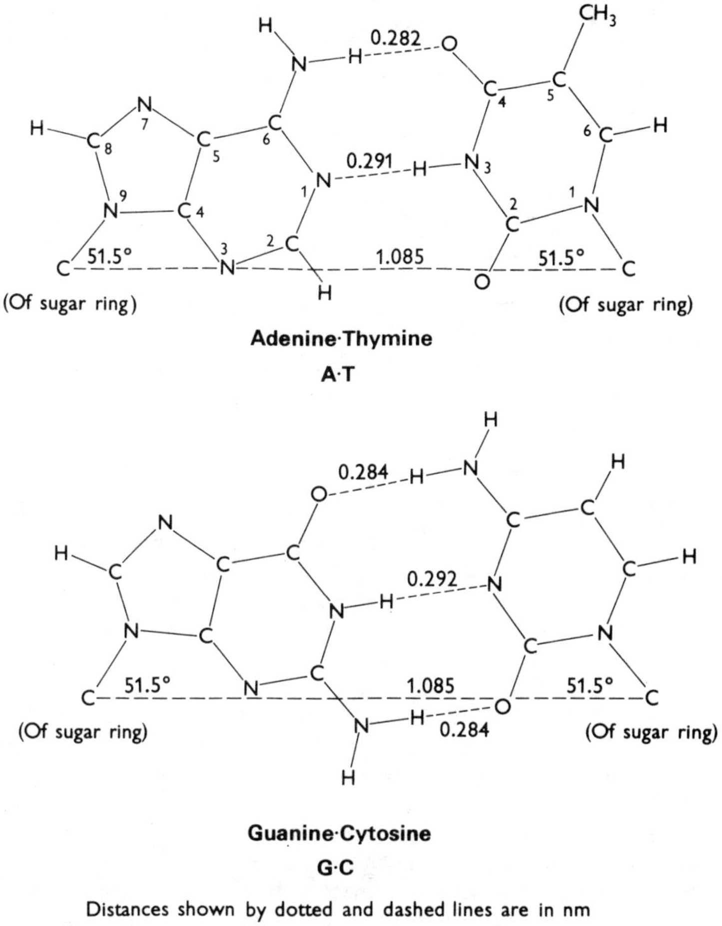

Adenine·Thymine

A·T

Guanine·Cytosine

G·C

Distances shown by dotted and dashed lines are in nm

Fig. 1–8 Standard (Watson-Crick) base pairs.

The biological significance of DNA structure

DNA, as we have already seen, is a rod-like structure consisting of two linear polydeoxyribonucleotides. In biological material DNA exists as extremely large molecules and is mostly located in eukaryotic nuclear chromosomes. Now we know that DNA is also found in mitochondria, chloroplasts and perhaps in other self-reproducing organelles. As an example of its immense size, the average molecular weight of the DNA in one of the 46 human chromosomes is 7.4×10^{10} daltons. So far it has not proved possible to isolate intact molecules of eukaryotic nuclear DNA. The long DNA molecules are easily broken by shearing forces but pieces of DNA up to 2 mm long, with a molecular weight of about 5×10^9 daltons, have been isolated. The prokaryotic DNA is also difficult to handle, but careful experiments have enabled the isolation of the intact total DNA molecule from *Escherichia coli*. This DNA (mol. wt. 2.5×10^9 daltons) constitutes the entire bacterial chromosome. This amount of DNA is sufficient to contain about 3000 to 4000 different genes (each gene being a stretch of DNA comprising 500 to 1000 base pairs) and in addition some base sequences which do not code for any product but which instead have regulatory or signalling functions.

A model for the genetic material has to account for two separate functions. It must have the potential of self-duplication and of giving rise to transformation into protein structures which direct the cellular metabolic reactions. The pertinence of the double helical model in self-duplication is readily seen and did not escape the notice of the originators. If the strands of DNA come apart in an environment containing the appropriate enzymes, deoxyribonucleotide subunits could be joined up on the separated chains so that the order of nucleotides is dictated by building to maintain complementary base pairing and so produce two DNA double helices from the original one. Indeed this is what happens in nature although we still do not know the full details. What has been determined experimentally is that the DNA molecule is duplicated during cell division so that one copy of the original molecule ends up in each of the two daughter cells. Moreover, it is known that one of the original DNA strands together with one new strand ends up in each daugher cell, thus confirming Watson and Crick's original proposal on how the model could be duplicated. This mechanism of duplication is called semi-conservative replication.

Until 1979, the DNA double helix was thought always to exist with a right-handed twist. Then Alex Rich and his colleagues demonstrated that small pieces of synthetic DNA could under special circumstances form a left handed helix called Z-DNA. In 1981, experiments using antibodies to Z-DNA have shown that Z-DNA can be found at some points of eukaryotic chromosomes. The current speculation is that left-handed DNA may have an important role in controlling the expression of genes in complex organisms.

The expression of genetic information in the cell will be described in Chapters 2 and 3. Here we need only say that the information content of the repeating phosphodiester-deoxyribose backbone must be nil so that we must consider the arrangement or order of the bases (i.e. the linear sequence of nucleotides) for the storage of genetic information. In other words the genetic information is encoded in the DNA sequence.

2 Elucidation of the Genetic Code

2.1 The Central Dogma

A gene is defined in general terms as the smallest unit of genetic information encoded in the genetic material. Thus a gene can code for other nucleic acids or proteins. In this chapter we shall discuss genes which code for polypeptide chains (proteins). In biochemical terminology such a piece of DNA coding for a polypeptide can also be called a cistron. The data that the order of amino acids in a protein was dictated by the order of nucleotides in a gene was generally accepted for several years before its experimental verification and was called the *sequence hypothesis*.

A one-way flow of the genetic information from nucleic acid to protein was assumed at an early stage by Francis Crick who formulated the Central Dogma of Molecular Biology based on this assumption. The discovery of an RNA intermediate in the flow of information from DNA to protein allowed him to state in 1958 the Central Dogma (see Fig. 2–1). The solid arrows show the main

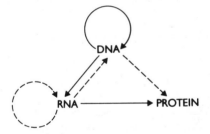

Fig. 2–1 The Central Dogma. Solid arrows represent general transfers whereas dotted arrows are special transfers. (Reproduced from CRICK. 1970.)

directional flows so that proteins do not contain genetic information for making nucleic acids. Recent discoveries that RNA provides information for making RNA or DNA and that DNA can specify protein in artificial systems are special cases which do not contradict the main directional flow.

2.2 Colinearity

Clearly the most logical way of envisaging the information transfer between gene and protein is by a parallel sequential read-out. This colinearity of gene and protein was assumed in the early 1960s but was not verified experimentally until 1964. Two approaches to the problem were successful at about the same time. One, used by Yanofsky and his colleagues (Table 1–1), involved the isolation and classification of mutant enzymes from related strains of the bacterium *Escherichia coli*. Mutants of the A protein of tryptophan synthetase were isolated and in each case the point in the polypeptide chain was determined where a single amino acid was different from that in the normal protein. The use of genetic methods permitted the 'mapping' of the mutations in the gene. A correlation of the order of mutations in the gene with the order of the amino acid differences confirmed their parallel sequential order (Fig. 2–2).

Experiment

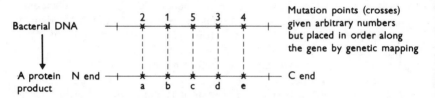

Gene for A protein of tryptophan synthetase between bars

Bacterial DNA

2 1 5 3 4

Mutation points (crosses) given arbitrary numbers but placed in order along the gene by genetic mapping

A protein   N end   C end
product   a   b   c   d   e

Amino acid substitutions due to various mutants

Result   The mutation points 2, 1, 5, 3, 4 are found to be in a sequential order corresponding to the amino acid alteration points a ⟶ e

Conclusion   The gene and the product are colinear

Fig. 2–2   See text for details.

Another approach was used by Brenner and his colleagues (Table 1–1). They were able to use a bacteriophage infected cell system in a manner which avoided the enormous amount of purification of proteins needed in the previous approach. Bacteriophage are bacterial viruses and usually consist only of nucleic acid genetic material encased in protein coats. Their genetic material can be DNA or RNA. Brenner's group used a special class of mutants of a DNA-containing bacteriophage called T4. This class of 'amber' mutants produces only fragments of polypeptide chains in bacteria. In particular, amber mutations in the gene carrying information for the coat protein introduce new points of termination during polypeptide synthesis so that only the N-terminal portions of the chain are made. Bacteria were thus infected with a series of amber mutants of the bacteriophage and the partial coat protein products were analysed to determine their sizes. The length of each fragment reflected the position of the termination point in the protein. The experience of bacterial geneticists indicated that the genetic map was linear and a genetic analysis of the amber mutants determined the order of the mutation points within the coat protein gene. When these points, placed in sequential order along the genetic map, were compared with the lengths of respective mutant polypeptide fragment products, the gene mutation points were found to be in the same order as the protein termination points, thus confirming the colinearity concept

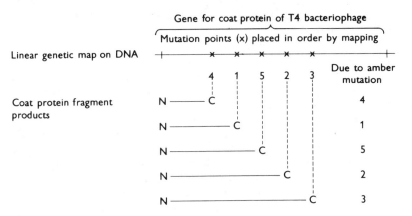

Gene for coat protein of T4 bacteriophage

Mutation points (x) placed in order by mapping

Linear genetic map on DNA

Coat protein fragment products

	4	1	5	2	3	Due to amber mutation
N —— C						4
N ———— C						1
N —————— C						5
N ———————— C						2
N —————————— C						3

Result   On analysis it was found that increasing lengths of coat protein fragments are produced all starting at the N-terminal end of the normal protein. The increasing lengths correspond to the increasing distances along the gene of the mutations which produce them.

Conclusion   Gene and protein are colinear

Fig. 2–3 See text for details.

(Fig. 2–3). In actual fact, this concept is not strictly true with respect to genes of higher organisms (see Fig. 2–6).

2.3   Properties of the code

Clearly the most direct and convincing way of deciphering the genetic code would be to compare the results of determining the nucleotide and amino acid sequences of a piece of DNA and its specific protein. Although this is possible nowadays, this problem was quite insoluble in the 1950s and the efforts of ingenious theoreticians who tried to make inspired guesses about the genetic code were hampered by a lack of experimental evidence. In 1959 any idea of solving the genetic code was out of the question. Yet the horizon began to brighten in 1960 with the implication that the newly discovered messenger RNA, a complement of genetic material, actually carried information for the direct specification of a protein sequence. Thus, in biochemical terms, the genetic code is the relationship of the nucleotide sequence of mRNA to the amino acid sequence of its relevant protein. The direct experimental elucidation of this relationship became a distinct possiblity through the work of Nirenberg and Matthaei in 1961 (Table 1–1). They were able to programme the synthesis of a polypeptide containing only one amino acid by the addition of a synthetic polyribonucleotide containing only one type of nucleotide to a bacterial cell extract. This vital step in the progress of our knowledge of protein biosynthesis is described more fully on p. 28.

At about the same time Ochoa and his collaborators (Table 1–1) constructed a bacterial cell-free system. The tremendous enthusiasm of both groups in a relatively competitive atmosphere was responsible for the rapid progress achieved over the next few years.

The other important contribution to our knowledge about the general nature of the genetic code came shortly afterwards from *in vivo* genetic studies by Crick and his collaborators (Table 1–1). This elegant genetic work was confirmed much later by *in vitro* biochemical studies with the advent of synthetic mRNA of chemically defined nucleotide sequences. Additional evidence of the general nature of the genetic code came from the amino acid analyses of human haemoglobins and from analyses of altered tobacco mosaic virus coat proteins where the changes were induced by chemical means. The general nature of the genetic code from all of these lines of evidence and from more recent confirmatory work can be summarized as follows:

(1) The genetic code is triplet in nature, i.e. a sequence of three nucleotides in mRNA specifies one amino acid in a protein. The group of three nucleotides used to specify an amino acid is called a codon or codeword. This property is strongly favoured by theory. The biochemists knew that mRNA contained only four nucleotides which had to be capable of coding the twenty common amino acids of bacterial proteins. It was obvious to them that out of possible arrangements, 4^1, 4^2 (4×4) and 4^3 ($4 \times 4 \times 4$) for the nucleotides as outlined in Fig. 2–4, $4 \times 4 \times 4$ ($= 64$) would be required to specify all the amino acids. Therefore, in the early 1960s, an arrangement of three nucleotides to specify an amino acid was assumed by the biochemists. This assumption dictated the next feature of the code which was found by the geneticists.

(2) The genetic code is degenerate, that is one amino acid (of which there are twenty) can be specified by more than one codon (of which there are sixty-four possibilities).

(3) The genetic code is non-overlapping and sequential. Thus the nucleotides are read off in groups of three (codons) in sequence from a fixed point (see Fig. 2–5 for the contrast with an overlapping code). The codons are contiguous so that under normal conditions of translation there are no gaps of meaningless codons. The genetic evidence strongly favours this third property in that a point mutation (a single base change) causes an alteration in only one amino acid of the protein sequence.

In 1976, the pioneering work of F. Sanger's group in Cambridge showed that the non-overlapping code is not universally true. For example, in the small DNA bacteriophage called ϕ X 174, overlapping of genes has been identified. This situation has also been found for several animal viruses. Whether this phenomenon is of widespread occurrence in organisms is not yet known. It may be relevant to systems where there is a compression of genetic information.

Another remarkable recent development arising from new rapid DNA sequencing methods has shown that an exception to the colinearity concept and sequential read out of genetic information can occur at least in eukaryotic cells and animal viruses. Inserted pieces of DNA called *introns* of unknown function and of different lengths can be shown to exist within structural genes.

Singlet Code (4 words)	Doublet Code (16 words)				Triplet Code (64 words)			
A	AA	AG	AC	AU	AAA	AAG	AAC	AAU
G	GA	GG	GC	GU	AGA	AGG	AGC	AGU
C	CA	CG	CC	CU	ACA	ACG	ACC	ACU
U	UA	UG	UC	UU	AUA	AUG	AUC	AUU
					GAA	GAG	GAC	GAU
					GGA	GGG	GGC	GGU
					GCA	GCG	GCC	GCU
					GUA	GUG	GUC	GUU
					CAA	CAG	CAC	CAU
					CGA	CGG	CGC	CGU
					CCA	CCG	CCC	CCU
					CUA	CUG	CUC	CUU
					UAA	UAG	UAC	UAU
					UGA	UGG	UGC	UGU
					UCA	UCG	UCC	UCU
					UUA	UUG	UUC	UUU

Fig. 2–4 Possible arrangements of bases in different types of code assuming that all codewords are the same length.

Furthermore, these inserted sequences become spliced out at the mRNA level before translation into protein as illustrated in Fig. 2–6. The functional gene pieces called *exons* code for protein regions, sometimes corresponding to three dimensional entities called domains.

2.4 Genetic experiments

The important genetic experiments performed by Crick's group can be briefly reviewed (Fig. 2–7) without going into detailed bacteriophage genetics. Various mutations in two linked cistrons (genes) defined in their position by genetic experiments, and called rIIA and tIIB cistrons of the wild type bacteriophage T4, were induced by chemical treatment with a derivative of the polyheterocyclic compound, acridine. This type of mutagen is thought to act by intercalation between the stacked bases of DNA. Although at that time the protein products from the rII cistrons were not isolated, the mutant proteins prevented the bacteriophage from growing in a normal host bacterium and so

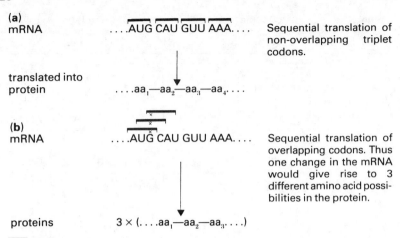

Fig. 2–5 Non-overlapping property of the genetic code (**a**) contrasted with an overlapping genetic code (**b**). Amino acids are abbreviated aa; the phosphates are usually omitted from the nucleic acid primary structure when the coding properties are discussed.

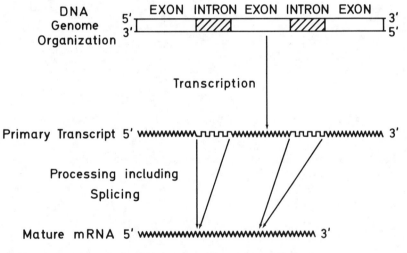

Fig. 2–6 Eukaryotic gene expression. See text for details.

its effect could be monitored. The genetic results could be rationalized by classifying the mutants produced by this mutagen into two types which appear to have either a base pair added or deleted in the DNA's nucleotide sequence. These mutants are clearly distinguishable from mutations which are considered to be base changes (by substitution). As shown in Fig. 2–7 it was found that when certain double mutants arose, genetically determined to be a combination of an addition type (+), and a deletion type (−), then the

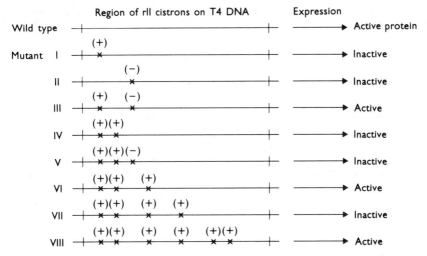

Fig. 2–7 Genetic evidence for a triplet code. Combination of acridine produced mutations showing the type of protein product in mutant bacteriophage grown on a standard host bacterium. (+) and (−) refer to differentiable classes of mutation which arise by the addition or deletion of a base pair in the gene.

For example if the wild type sequence of codons is

          ... AUG CAU GUU AUU ...

then a (+) mutation with an insertion at the arrow to alter the reading frame could be

                   (+)
                   ↓
          ... AUG CCA UGU UAU ...

and a (−) mutation with another alteration in reading frame, the following

                   (−)
                   ↑
          ... AUG CAU GUA UUU ...

Thus a combination of (+) and (−) restores the reading frame with a short piece of altered amino acid sequence not affecting the protein's function

             (+)   (−)
             ↑   ↑
       ... AUG CCA UGU AUU ...

bacteriophage behaved as a wild type (i.e. a pseudo wild type) in most cases. A restriction was imposed upon the recovery of wild type character by a double mutant; it happened only when the mutations were not too far apart on the genetic map. However, the fact that many combinations of a (+) (arbitrarily designated as an addition) and a (−) mutant realized a pseudo wild type bacteriophage indicated that most codons could be translated into amino acids (i.e. as meaningful or *sense* codons), and that triplets which could not be translated into amino acids (non-meaningful or *nonsense* codons) were relatively rare. This strongly supported the contention that the genetic code was degenerate. An extension of the properties of double mutants indicated that the reading frame for the protein specification started from a fixed point. This idea came from a rationalization of the properties of some rare untranslatable parts of the rII cistrons thus defined as *barriers*. When a (+) and (−) combination spanned a barrier the effect of adding and deleting a base pair did not give rise to active proteins, i.e. characterized by wild type bacteriophage. Yet, if the combination was on the same side of a barrier, normal translation occurred with a small segment of altered protein sequence being incorporated into the mutant bacteriophage. This hypothesis for the mechanism of action of acridine mutagens or other frame shifting mutagens (Fig. 2–7) has been experimentally supported by Streisinger and his colleagues (TERZAGHI *et al.* 1966) in experiments to show the polarity of reading of the genetic message *in vivo*. In these experiments (see section 2.12) the amino acid sequence of altered protein was actually determined and shown to be different between the double mutation points. Furthermore Streisinger's experiments allowed the verification of the assignment of a base pair addition to a (+) mutant. To determine the size of the coding ratio, Crick and his colleagues constructed bacteriophage containing multiple mutants with and without mixtures of the two types (+) and (−). When a bacteriophage was constructed from three (or a multiple of three) all (+) or all (−) mutations, pseudo wild types were obtained, indicating that the code is read in groups of three (or of multiples of three) nucleotides at a time (Fig. 2–7).

2.5   Biochemical experiments: the cell-free system

For several years prior to the cell-free experiments carried out by Nirenberg and Matthaei, research biochemists had been breaking open cells from various source materials and attempting to reassemble an active mixture of components for the synthesis of proteins. The methods of breakage included grinding with an abrasive material such as alumina or glass beads and forcing cells in solution or frozen through a very small hole under pressure. Radioactively labelled amino acids were added to these crude cell extracts and assays for newly synthesized radioactive proteins were devised. However, the incorporation of radioactive amino acids was never observed in amounts which signified the construction of an active cell-free system. NIRENBERG and MATTHAEI (1961) were able firstly to construct a relatively stable cell extract with intact components for the synthesis of proteins, and secondly to direct the synthesis of a protein-like product with a synthetic polyribonucleotide which acted as mRNA. Thus, they showed biochemically that mRNA was directly translated into protein and opened the way for an experimental investigation to elucidate the genetic code. During the rapid expansion of this field, experimental results identified the components and substantiated the scheme for the mechanism of protein biosynthesis. Furthermore, one of Nirenberg and Matthaei's original experiments showed that DNA controls protein synthesis, substantiating the idea that mRNA is an intermediate step in the transfer of information.

They found that by the addition of a *sulphydryl agent* (i.e. a reagent such as 2-mercaptoethanol containing a reactive SH-group to prevent the oxidation and linking up of such active groups in the cell extract proteins) to an extract of broken bacterial (*Escherichia coli*) cells they could stabilize the components involved in protein synthesis. If they destroyed the bacterial DNA by the addition of the enzyme pancreatic deoxyribonuclease (DNase) (Fig. 2–8), the indigenous capacity of the cell extract for incorporating amino acids was decreased. Later they and other workers constructed a cell extract which incorporated radioactively labelled amino acids into newly synthesized protein dependent upon added (exogenous) DNA such as bacteriophage DNA. Another bonus was obtained by the addition of DNase during cell breakage. The cell extracts in the absence of gelatinous DNA were much easier to manipulate and separate into components. The cell extract which was used to construct the first cell-free system was simply obtained by breaking cells, degrading the DNA, and removing the cell debris including cell walls by centrifugation. The active suspension of components was called an *S–30*

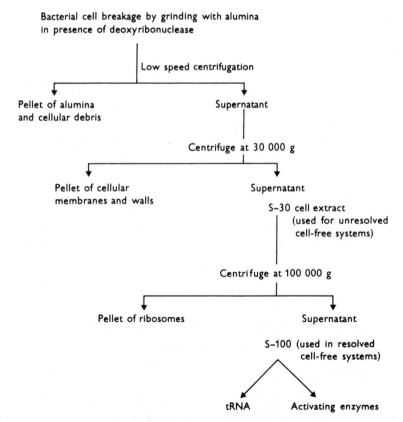

Fig. 2–8   Preparation of the cell-free system and its components.

because of the centrifugal forces needed for its preparation. This means that the cell extract S–30 supernatant was the result of spinning in a refrigerated high-speed centrifuge so that a force of 30 000 × force of gravity (g) had been applied to it. All the manipulations for the preparation in Fig. 2–8 were carried out with the material kept cold (0–4°C) to help stabilize them.

At first Nirenberg and Matthaei worked non-stop from the preparation of their bacterial cell extracts through to their experiment to avoid the inactivation of their working material. Later their technology was improved and the extracts stabilized so that they could be frozen in dry ice or in liquid nitrogen and stored for months without too much inactivation. The spectacular result came when they added the standard set of twenty amino acids, each of which was radioactively labelled in turn, to their cell extract which was fortified with inorganic salts, a buffer and an ATP generating system to maintain energy requirements. When they added a synthetic polyribonucleotide (for structures and abbreviation see section 1.5) containing only uridylate residues (poly U) and in the presence of radioactive phenylalanine, the amount of radioactive polypeptide formed increased greatly (Fig. 2–9). Since poly U was found to code for polyphenylalanine, UUU would be the codeword for phenylalanine if the triplet code were correct. The radioactive product was collected as a precipitate on nitrocellulose filters (see section 2.6). In point of fact, the most favourable increase of incorporation over background (no added polyribonucleotide) could be obtained by exhausting the S–30 extract of intact mRNA. This was achieved by a system of preincubation and then dialysis to get rid of small waste products before the main incubation was performed. This may well be looked upon as the birth of the *in vitro* or cell-free system.

Of course, it was a favourable meeting of events since the earlier discovery of the enzyme, polynucleotide phosphorylase, permitted the synthesis of polyribonucleotides such as poly U (Fig. 2–10). Indeed, Ochoa (a Nobel prize winner for RNA synthesis) and his co-workers, who first isolated polynucleotide phosphorylase from *Azobacter vinelandii*, were not far behind Nirenberg and Matthaei in constructing a cell-free system which was active for protein biosynthesis. Both groups of workers constructed similar cell extracts but one clear difference was in the compounds used to regenerate the energy provider, ATP.

[^{14}C]-Phe

radioactive phenylalanine

E. coli	Components: S–30 made as in Fig. 2–8 mercaptoethanol,
Cell-free	magnesium and potassium ions, Tris buffer pH 7.8, ATP,
system	phosphoenol pyruvate, pyruvate kinase and poly U.

[^{14}C]-poly Phe

Fig. 2–9 Poly U makes poly Phe in a cell-free system. Tris is an abbreviation for trishydroxymethylaminomethane popularly used as a buffer for biological systems because of its harmlessness. ATP is adenosine 5'-triphosphate.

Homopolymer

$$n \times ppU \quad \xrightarrow[\text{phosphorylase}]{\text{Polynucleotide}} \quad (pU)_n + n \times p$$

Uridine   poly U   inorganic
diphosphate   phosphate

Copolymer

$$m \times ppC + n \times ppU \quad \xrightarrow[\text{phosphorylase}]{\text{Polynucleotide}} \quad (pC)_m \, (pU)_n + (m + n) \times p$$

Cytidine   poly (C, U)
diphosphate

n and m are integers

Fig. 2–10   Synthesis of polyribonucleotides for use as mRNA.

In a fairly competitive spirit, both groups proceeded to synthesize polyribonucleotides containing more than one base component to see whether other amino acids could be incorporated. However, before proceeding with this phase of the elucidation of the genetic code, the implication of two extremely important components of protein biosynthesis in this early work must be described.

2.6   Resolution of the cell-free system

Since the S–30 cell extract contained so many unknown materials, it was clear that it would greatly help our understanding of protein biosynthesis if the S–30 extract could be resolved into active components. Firstly, further high-speed ultracentrifugation allowed the isolation of nucleoprotein ribosomes (Fig. 2–8) which, when combined with their supernatant S–100, gave an active amino acid incorporating system. Naturally enough the combination of resolved components was not as active or as stable as the S–30 system. Now that isolated ribosomes could be employed in cell-free protein synthesis, Nirenberg and his group soon completed the experiment to show that aminoacyl-transfer RNA (in those days called soluble RNA) is an intermediate in protein biosynthesis (Fig. 2–11). This experiment needed the resolved components, S–100 and ribosomes. The tRNA was isolated independently from bacterial cells and amino acids attached by a mixture of aminoacyl-tRNA synthetases (amino acid activating enzymes, see Chapter 3) isolated from the S–100 extract. Of the attached amino acids the phenylalanine (Phe) was radioactively labelled. When the radioactive phenylalanyl-tRNA was added to the cell-free system in the presence of poly U, poly Phe was synthesized showing that Phe-tRNA was an intermediate in its synthesis.

At this stage it is important to realize how the synthesis of protein was assayed. Nirenberg and Matthaei's assay was an improvement on the crude methods used previously. They introduced the Millipore filter to this field of

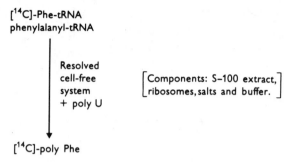

Fig. 2–11   tRNA is an intermediate in protein biosynthesis.

biochemistry and encouraged its use in simplifying many procedures. Previously proteins had been precipitated, stirred into suspensions on aluminium planchets which were then dried and their radioactivity counted in a gas-flow Geiger counter. The Millipore Company manufactures nitrocellulose filters with 0.45 μm diameter holes which easily retain fine protein precipitates resulting from using dilute trichloroacetic acid (TCA) as a precipitant. The precipitate was finely distributed over the surface of the filter and, when dried, could be assayed for radioactivity either by gas flow counting or by liquid scintillation counting. However, in this early work the use of nitrocellulose filters was not absolutely necessary; filters made of other suitable materials such as glass fibre could be used.

The aminoacyl-tRNA was also precipitated by TCA so high backgrounds in the resolved systems had to be avoided by heating the TCA-precipitated material at 90°C for 5 min. This treatment hydrolysed the amino acid from the tRNA so that the radioactivity in the precipitate came only from that incorporated into polypeptide. Hence at the end of an incubation (at 37°C) of the cell-free system components, the mixture was adjusted to 5% (mass/vol.) in TCA and heated before filtering off the radioactive protein. Aminoacyl-tRNA is still readily assayed by precipitation with cold TCA.

2.7   Use of synthetic polyribonucleotides

Although the resolved cell-free system (i.e. where ribosomes and S–100 are recombined) was very important in defining the components of protein biosynthesis, the S–30 cell extract was most importantly used to acquire new information about the genetic code. The early progress, as has already been mentioned, was due to work led by Nirenberg and by Ochoa. Polyribonucleotides of differing base composition were synthesized using the enzyme polynucleotide phosphorylase (Fig. 2–10) and these were tested in the cell-free system to determine which amino acids they incorporated into hot TCA-precipitable material. A convenient source of polynucleotide phosphorylase was found to be *Micrococcus lysodeikticus* and this helped in the proliferation of the available synthetic mRNAs.

The early cell-free system results were not very reliable and in retrospect it is remarkable how accurate they were. Obvious selection was made in the assay system for material insoluble in TCA. Had the polypeptide product been soluble it would have been missed. In fact, this occurred when poly A was tested for messenger activity. The product polylysine was missed for a year, before a better method of assaying was adopted by Ochoa's group. They used an acidified sodium tungstate solution as precipitant. In addition, there were many reasons why the various synthetic polyribonucleotides could act with differing efficiencies irrespective of their base sequence. Their efficiency depends on their molecular weight (a function of their length) and secondary structure (a reflection of their base composition).

In the early experiments with copolymers (polymers containing more than one type of nucleotide) as mRNA, only U-containing polymers were used because of the ready insolubility of phenylalanine-containing polypeptides in hot TCA. The nucleotide sequence obtained from using differing input ratios of nucleoside diphosphates during the polynucleotide's synthesis was assumed to be random. The continuing accuracy of the resulting assignment justified this assumption. Thus when the ratio of amounts of radioactive amino acids incorporated into polypeptide was calculated and compared with base ratios of the polynucleotide used as mRNA in the cell-free system, triplet codons were assigned to amino acids. These experiments permitted the assignment of codons only in terms of base composition, not sequence except for the homopolymer incorporation results, for example UUU was clearly the codon for Phe (phenylalanine). The possible inaccuracy from assuming that the base ratio of the copolymer depended upon the input ratio of the nucleoside diphosphates during the enzymic synthesis was realized when the base ratios were actually determined. Comparison of the analytical results, especially for guanine-containing copolymers, differed significantly from the assumed base ratios. After this discovery all the laboratories concerned adopted the practice of basing their calculations upon analytical data on the copolymers.

It was soon realized from analytical results of naturally occurring mRNA that non U-containing codons must occur. Workers in Europe were the first to show that a non U-containing copolymer, poly (C, A) (a polyribonucleotide containing cytidylate and adenylate in random order) could stimulate the synthesis of a polypeptide; in this case containing mainly proline and threonine. Following this result the two main groups in the field, after a period of intensive effort making use of a large variety of copolymers containing combinations of up to all four nucleotides, showed that most of the possible codons could be assigned to amino acids. This effort could be looked upon as making many entries in our dictionary of codewords used in encoding information in mRNA but only the composition, not the sequence, of codewords was known.

Although the deficiencies of the copolymer-stimulated system were apparent and it was realized that minor effects should be treated with scepticism, for the most part these assignments were correct. One error obvious to us now was the codon AAU for lysine.

Some additional results were achieved by using polyribonucleotides containing ends of defined nucleotide sequences. Polynucleotide phosphorylase can build units of a different nucleoside diphosphate on to a polynucleotide primer and so achieve different end sequences. Indeed Ochoa's group decided the direction of reading of the mRNA for protein synthesis (see section 2.12) by using a type of block polymer. Although the data obtained with such a polymer containing a length of 20 A nucleotides and 1 C, (poly A_{20} C) were not absolutely convincing, the conclusion was subsequently confirmed by other methods, most prettily by Streisinger and his colleagues, who determined amino acid sequences within a mutated piece of the T4 bacteriophage enzyme, lysozyme (see section 2.12). Part of the problem using block polymers (polymers containing lengths of different nucleotides) such as poly $(A_{20}$ C) was due to the presence of nucleases (nucleic acid degradative enzymes) in the *E. coli* cell extracts. Ochoa's group overcame this problem by constructing their cell-free system from a combination of carefully washed *E. coli* ribosomes and a ribosomal supernatant from a class of nuclease-free bacteria – *Lactobacillus arabinosus*.

During the copolymer work a strange gap persisted in the experimental basis of our ideas about cell-free protein synthesis. Everyone assumed that polypeptides were made in the cell-free incorporation of amino acids because of the solubility and radioactive properties of the products. However, more critical workers began to require further confirmation that polypeptides were made and, in 1963, several groups took pains to isolate the polypeptide product caused by polyribonucleotide messengers and to analyse it by breaking it down into its constituent amino acids. Polylysine was isolated and characterized after it was synthesized in a cell-free system programmed by poly A, then it was hydrolysed by treatment with acids and by the proteolytic enzyme trypsin. The small chains of oligolysine and free lysine identified in the hydrolysate showed that only polylysine had been formed and that the lysine units were joined by peptide bonds.

This work was part of a larger two-stage experiment in which the author took part which attempted to construct a model system for studying the control of protein synthesis by DNA. It was known that exogenous DNA added to a cell-free system devoid of DNA would stimulate the incorporation of amino acids. Furthermore the cell-free system could be fractionated so that the enzyme for transcribing DNA, RNA polymerase, could be used in a first stage to synthesize a polyribonucleotide complement of added DNA. Then in a second stage incubation, which included the addition of a cell extract (supplemented with buffered inorganic salts and amino acids) to the first stage reaction mixture, a polypeptide product was synthesized from genetic information encoded in the DNA or model DNA added to the first stage. In trying to set up a simplified model system to further our knowledge about the transfer of genetic information, we decided to synthesize chemically a short piece of polydeoxynucleotide (i.e. an oligodeoxynucleotide) according to the well established procedures worked out by Khorana and his colleagues (Table 1–1). The first simple model DNA chosen contained only one type of base. We made a series of oligodeoxythymidylates, oligo dT, of increasing chain length up to 14 units long. The transcription of such material with the enzyme RNA polymerase was studied at the same time by other laboratories including Khorana's. We were able to construct a two-stage system for making polypeptide from an oligodeoxynucleotide and, as already stated, used this system to verify that peptide bonds were formed in the cell-free system. An interesting feature of the first stage reaction was that the poly A made was much longer than the template oligo dT. This feature, thought to be due to a slipping and copying mechanism (reiterative copying) during transcription by the RNA polymerase, has been used by Khorana's group to make polyribonucleotide copolymers of chemically defined sequences (section 2.9).

In 1963 it was clear that the use of random copolymers as mRNA would not give the sequences of nucleotides within codons. Further progress in this field needed the isolation of a natural or synthetic mRNA of known nucleotide sequence whose polypeptide product in a cell-free system could be analysed. The chemical synthesis of oligodeoxynucleotides of known nucleotide sequence with the possibility of transcribing these into synthetic mRNA of known sequence was the first step in this direction. Khorana's group has used this approach to confirm and establish experimentally most of the features of protein biosynthesis related to the transfer of genetic information although they were not the first to elucidate nucleotide sequences of codons.

2.8   The triplet binding assay

Progress in elucidation came in another major breakthrough by Nirenberg's group at a time when the technology involved was so complicated as to dampen the enthusiasm of biochemists. The simple elegance of a new method, the triplet binding assay, easily persuaded Nirenberg and his colleagues to drop the earlier technique which needed the laborious chemical synthesis of oligodeoxynucleotides. Early in 1964 it was known that synthetic polyribonucleotides stimulated the binding of aminoacyl-tRNA to ribosomes. Indeed, the complex of these components could be located in an analysis of the product by ultracentrifugation through a density gradient of sucrose solution, but this is a very tedious method and to do many assays in this way would be out of the question. Nirenberg and his colleagues investigated the possibility of isolating the aminoacyl-tRNA coupled with ribosomes and mRNA by a quicker method. The Millipore filter made of nitrocellulose again came to the rescue, this time in a spectacular manner. Ribosomes were found to adsorb to these filters and so did some polyribonucleotides, including poly U alone. However, aminoacyl-tRNA tagged by a radioactive label on the amino acid passed through the filter unless it was bound to the ribosome. In particular NIRENBERG and LEDER (1964) were able to induce the binding of Phe-tRNA to ribosomes by poly U and not by other homopolymers, the binding complex being adsorbed on a nitrocellulose filter. The importance of this type of complex as a coded intermediate in protein biosynthesis was confirmed using polyribonucleotide copolymers as synthetic mRNA. It was shown that the species of aminoacyl-tRNAs which were bound to ribosomes by the copolymers were derived from the same amino acids which were incorporated by the copolymers in the cell-free system.

The components of their reaction mixture (Table 2–1) suggested that the reaction was non-enzymic; no cell extract supernatant enzymes were involved. (It was later shown that this is not strictly so – see Chapter 3.) Furthermore, short lengths of poly U even as short as the triplet, UpUpU, stimulated the formation of the complex.

There is a conventional distinction between triplets and codons and their abbreviations. A triplet now usually describes a short piece of nucleic acid containing three bases as used in the elucidation of the genetic code. These structures, abbreviated for example as UpUpU, are not trinucleotides because they are one phosphate short, and are correctly called trinucleoside diphosphates. This type of triplet was used in the triplet binding assay. However, when transmission of genetic information and coding properties are described, for convenience the phosphates are omitted as the information is encoded only in the order of the bases. Thus the codon for phenylalanine is called UUU which is derived from the experimental observation using the triplet UpUpU.

In addition to confirming the triplet nature of the genetic code directly, Nirenberg and his colleagues realized that it was a very much easier problem to construct trinucleotides or equally active trinucleoside diphosphates of known sequences rather than synthetic mRNA. Their optimism proved justified and their work, which started with successful use of the first heterobase triplet GpUpU (Table 2–1) allowing the assignment of GUU to valine, rapidly progressed so that about fifty out of the sixty-four codon sequences were assigned, directly and convincingly, using this method.

I say convincingly, even though in some cases the binding assay gives unreliable results on account of the small effect observed. Some increase in the effect is seen when a purified tRNA species is used. This feature stimulated the fractionation of mixed bacterial tRNAs so that as many tRNAs as possible, each specifically chargeable with one amino acid, could be tested in the binding

Table 2–1   Components of the triplet binding assay.

In 0.05 ml incubated at 20°C for 25 min

---

6 n mol GpUpU
0.5 A_{260} units of aa-tRNA labelled with ^{14}C-Val
1 A_{260} units of ribosomes
0.02M Mg acetate
0.05M K chloride
0.1M Tris acetate pH 7.2

Nucleic acids and ribosomes are conveniently quantitated in ultraviolet absorption units at a wavelength of 260 nm. Tris is an abbreviation for a fairly harmless buffer used in biochemistry; n mol is a standard abbreviation for nano moles or $10^{-9} \times$ moles.

assay. It should be pointed out that fractionation of tRNA species has in some cases yielded several tRNA species each of which is chargeable with the same amino acid yet is bound by a different triplet, giving direct proof of the degeneracy of the genetic code. In practice the tRNA species were seldom purified to homogeneity since it was found that partial purification would suffice to show an effect. When the binding complexes with available triplets were still not stable enough to be detected by this method, codons could be determined and others confirmed by means of the cell-free system programmed by synthetic mRNA of defined sequences as used in Khorana's laboratory (see section 2.9). Another possible method is biochemical analysis of changed polypeptides produced from *in vivo* mutagenic events. The elegant experiments of Streisinger's group to obtain *in vivo* information on the polarity of reading of the genetic message are described in section 2.12.

This progress in the elucidation of the genetic code had important side effects in the stimulus afforded to the synthesis of oligonucleotides and to the fractionation of tRNA species. A search for quicker methods of synthesizing triplets led to the widespread use of polynucleotide phosphorylase either under conditions where the enzymic reaction was slowed down so that a product could be isolated after the addition of a few nucleotide residues, or where polymers made on a dinucleoside phosphate primer were degraded from the 3'-end (see section 1.5) by long equilibration in the presence of the enzyme, so that the resultant oligoribonucleotide reflected the input ratio of dinucleoside phosphate to nucleoside diphosphate. Progress has been made in the commercial field so that all sixteen possible dinucleoside phosphates are available. In 1964 they were obtained in the laboratory by chemical synthesis or by degrading RNA. Of course, the most readily applicable method for all sixty-four triplets is chemical synthesis, but this method of approach could only reasonably be attempted in a few laboratories. Indeed, the need for trinucleoside diphosphates in elucidating the genetic code provided a stimulus for the organic chemists in Khorana's laboratory to synthesize all the possible sixty-four trinucleoside diphosphates in a matter of a year. This great feat helped in the rapid elucidation of the genetic code yet the biochemists managed to synthesize most of the triplets more quickly since they were not concerned with the preparation of intermediates containing possible interferingly reactive groups protected chemically and did not require so much material at each stage in the overall synthesis.

2.9   Use of synthetic polyribonucleotides of known sequence

The best experimental evidence for the assignment of many of the codon sequences has come from the workers in Khorana's laboratory using synthetic mRNA of known nucleotide sequences. They solved the problem of relating the nucleotide sequence of a synthetic mRNA directly to the amino acid sequence of its polypeptide product by an admirable combination of organic chemical synthesis and enzymic synthesis. First they used the reiterative copying mechanism of the enzyme DNA polymerase to make long chain polydeoxyribonucleotides by extending chemically synthesized double helical

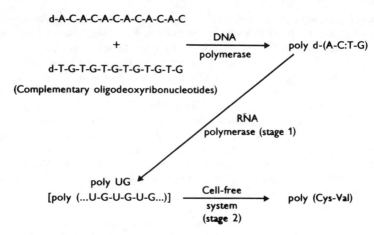

Fig. 2–12 See text for details.

oligodeoxyribonucleotide of about twelve units and of known sequence as in the scheme outlined in Fig. 2–12. This chain extension is probably accomplished by one chain slipping relative to the other (in this case by 2 units) yet maintaining base pairing, thus exposing ends on opposite chains which can be filled in or repaired to maintain the double helix by the enzyme incorporating units from the substrate nucleoside triphosphates. Once the long chain double stranded synthetic DNA is made, this synthetic gene of known sequence can be used many times since it is easily recovered from the cell-free system. A polypeptide, whose amino acid sequence is determined by analysis of its radioactive constituents, is synthesized in a two-stage reaction in which the cell-free extract is added (such as an S–30 extract) to the mRNA synthesized (by RNA polymerase copying the synthetic DNA) in the first stage. The scheme makes it clear that to make the specified polyribonucleotide, poly (U–G) containing U and G in a repeating sequence only the triphosphates containing U and G are added to the reaction to ensure the copying of only one strand of the DNA. Since the polypeptide in the example shown contained a repeating cysteine-valine sequence and UGU was known to be a codon for cysteine from binding studies, it was possible to assign the other possible codon, GUG, to valine from the repeating UG sequence, i.e. U–G–U–G–U–G, corresponding to cysteine-valine. This assignment was made in fact before it was confirmed by the triplet binding assay. The synthesis of defined polyribonucleotide sequences has been extended by Khorana's group to include all the types shown in Fig. 2–13.

2.10   Characteristics of the complete code

By the end of 1966 the elucidation of the genetic code was complete (CRICK 1968). The full assignment of codon sequences is shown in Table 2–2. Although the laboratories of Nirenberg and Khorana were largely responsible for allocating codon sequences to amino acids (for which the leaders gained a share

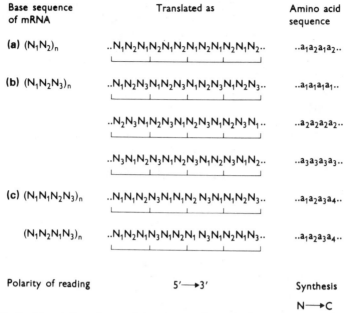

Base sequence of mRNA	Translated as	Amino acid sequence

Fig. 2–13 Translation of repeating sequences in synthetic mRNAs by the cell-free system. The synthetic mRNAs contain repeating nucleotide sequences in groups of two, three or four ((**a**), (**b**) and (**c**) respectively). When these mRNAs are translated in the cell-free system the amino acid sequences at the right are obtained. a = amino acid.

in a Nobel prize), there were also many useful contributions by other research workers using oligonucleotides of defined sequences, in both the binding assay and the cell-free system.

As predicted by the geneticists, there are very few nonsense codons. The codons UGA, UAG and UAA are signals for polypeptide chain termination. These codons do not normally signify amino acids but can be decoded by specially altered tRNAs called suppressor tRNAs because they suppress normal polypeptide termination. There is good evidence that codons UAG and UAA signify chain termination in eukaryotic cells as well as bacterial ones, but whether UGA is a ter codon in eukaryotes is unclear. It is known that it sometimes stands for Trp.

With respect to suppression of ter signals, an interesting and possibly more effective use of genetic information in eukaryotic cells has recently been proposed. Natural suppression of chain termination signals might occur to elongate proteins and therefore provide two gene products from the same gene (see section 4.4).

Polypeptide chain initiation differs strikingly from termination in that the codons assigned for initiation also signify an amino acid when the codon occurs internally in the reading of mRNA. The results from *in vitro* experiments identify AUG and GUG as initiation codons whereas they also signify methionine and valine, respectively, for internal translation. Whether the codon alone is a complete initiation signal will be discussed in Chapter 4. Both AUG and GUG have been identified as initiation codons in many natural mRNA sequences. Termination signals are not usually decoded by a special tRNA whereas initiation signals are by a special type of initiator tRNA, formylmethionyl-tRNA in bacteria and mitochondria, and a special non-formylated methionyl-tRNA in eukaryotic cytoplasm.

Although the genetic code was elucidated by *in vitro* experiments, the correctness of the codon sequences was confirmed first by analyses of mutant proteins and in recent years by actual sequencing of natural mRNAs or their genes for known protein sequences. Where amino acid sequence analyses of mutant proteins have identified amino acid replacements, these amino acid codons are, as expected, related by a single base change from a single point mutation in the gene.

Table 2–2 The genetic code. An arrangement of the codon assignment first due to F. H. C. Crick.

SECOND LETTER

	U		C		A		G		
U	UUU UUC	Phe	UCU UCC		UAU UAC	Tyr	UGU UGC	Cys	U C
	UUA UUG	Leu	UGA UCG	Ser	UAA UAG	Ter	UGA UGG	Ter Trp	A G
C	CUU CUC		CCU CCC		CAU CAC	His	CGU CGC		U C
	CUA CUG	Leu	CCA CCG	Pro	CAA CAG	Gln	CGA CGG	Arg	A G
A	AUU AUC	Ile	ACU ACC		AAU AAC	Asn	AGU AGC	Ser	U C
	AUA AUG	Met	ACA ACG	Thr	AAA AAG	Lys	AGA AGG	Arg	A G
G	GUU GUC		GCU GCC		GAU GAC	Asp	GGU GGC		U C
	GUA GUG	Val	GCA GCG	Ala	GAA GAG	Glu	GGA GGG	Gly	A G

Ter indicates a termination (or nonsense) codon

2.11   Wobble hypothesis

The most helpful rationalization of patterns occurring in the genetic code dictionary was made by CRICK (1966) in his 'Wobble Theory' (Table 2–3). This is derived from a knowledge of how the tRNA is concerned in the transfer of information from mRNA to protein.

Table 2–3 The wobble hypothesis. Predicted base pairing for the third position of a codon with tRNA.

3rd position of codon		1st position of anticodon
A or G	← decoded by →	U
G	← →	C
U	← →	A
U or C	← →	G
U, C or A	← →	I

I is inosine, a rare base found in yeast tRNA. It can be formed by deamination of the 6–NH$_2$ group of A.

An elegant experiment by Chapeville and his colleagues, in the days when random copolymers were used as synthetic mRNA, showed that once the amino acid was attached to a tRNA, it could be converted into another amino acid that was still coded by the same polynucleotide (see Chapter 3). Thus the tRNA was the recognition feature for transferring coding information. It was predicted that the tRNA would recognize the mRNA by the same base pairing mechanism that operates in double stranded DNA (see section 1.6). Since the code is triplet in nature a sequence of three nucleotides in the tRNA might be its decoding feature. This prediction turned out to be correct and the position of the three nucleotides, the anticodon, in the sequence of tRNA has been well established (see Chapter 3).

Crick noticed that amino acid codons could be grouped in sets according to differences in only the base of the third position of the codon. For example, UU (U or C) coded for phenylalanine, UU (A or G) coded leucine and it was thought that in a yeast system GC (U, C or A) coded alanine. A quick incursion into the realms of model building enabled him to suggest that certain non-Watson–Crick base pairs (i.e. other than G.C and A.U) were possible with only slight distortion of a double helix. He proposed that such a slight distortion could be possible during the interaction of mRNA with tRNA and new rules of pairing of the third base (3′-end) of the codon with the first base (5′-end) of the anticodon were devised as shown in Table 2–3. (Notice that codon and anticodon are complementary and antiparallel as in the DNA double helix.) Even the first two tRNA sequences published supported the hypothesis: they gave the anticodon assignments IGC for alanine tRNA and IGA for serine tRNA. I is the abbreviation for inosine, a rare nucleoside found in tRNA with some coding characteristics of G but it also base pairs with A. Chemically it can be formed from A by converting the N–6 amino group into a keto (C = O) group. Although these tRNAs came from yeast, the Ala-tRNA from yeast was coded on bacterial ribosomes in the triplet binding assay as GC (U, C or A) consistent with the anticodon IGC. Determination of nucleotide sequences of anticodons of other bacterial tRNAs show a similar pattern of coding relationships consistent with the Wobble Hypothesis. An important aspect of the hypothesis is that it readily explains the existence of a high degree of *ambiguity* within the multiple species of tRNA, i.e. having more than one codon for the same tRNA. For example, two species of phenylalanyl-tRNA could have anticodons GAA and AAA with codons UU (U or C) and UUU respectively.

In recent years, several workers have questioned the complete validity of wobble, based chiefly on results from *in vitro* studies. Such studies showed that some alternative 3rd base codon 1st base anticodon decoding interactions apparently were possible. These observations led U. Lagerkvist to propose that an actual decoding interaction does not occur in triplet terms so that in cases where 4 codons differing only in the third position code for one amino acid, the reading of bases in a decoding interaction is only 2 out of 3. However, there is good evidence from *in vivo* studies using yeast and *E. coli* systems that 3 out of 3 reading does indeed occur. One rationalization is that the *in vitro* studies show up more possibilities than actually occur *in vivo*.

2.12 Polarity of reading the genetic message

A striking proof of the direction in which mRNA is translated was afforded by the *in vivo* experiments of Streisinger's group (TERZAGHI *et al.* 1966). The amino acid sequences of similar parts of two types of the enzyme lysozyme produced by the bacteriophage T4 were determined (Fig. 2–14). The sequences corresponded to part of the gene where a base deletion and insertion had occurred in the production of a phase shifted double mutant by treatment with acridine mutagens (see section 2.4). When the nucleotide sequences of the mRNAs are written down from a knowledge of the genetic code, the sequences can be rationalized in terms of the base deletion and insertion shown, and it was possible to assign CUU and UUA to leucine from this examination. Furthermore, if the mRNA sequences were written in the opposite direction with respect to the amino acid sequence, the codon assignments could not be correlated as above. Thus the mRNA was translated in the direction from the 5'-end to the 3'-end to make protein from the N-terminal to the C-terminal end. This polarity of reading of the genetic message was confirmed by the experiments illustrated in Fig. 2–13.

2.13 Accuracy in translation of the genetic code

There are at least two places in the translation of the mRNA where errors are possible. Firstly, an amino acid must be activated for attachment to a specific tRNA. Possibly a slightly different amino acid occasionally becomes attached to the specific tRNA. Secondly, the aminoacyl-tRNA must decode a specific codon of the mRNA with a small possiblity of interference from other aminoacyl-tRNA species.

There must be a limit to the capacity of any biological system to discriminate between similar structures or surfaces so there must be a finite error in translation as well as in transcription. Very little is known about any possible

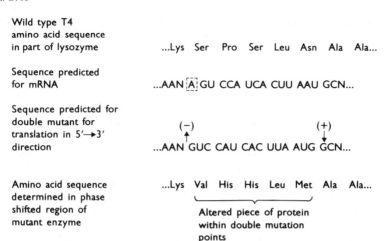

Conclusion

Therefore polarity was as below

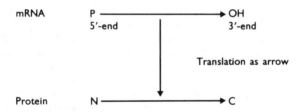

Fig. 2–14 Polarity of reading of the genetic message from the structure of a phase shifted piece of bacteriophage lysozyme. (N is an unspecified base; (−), base deletion; (+), base insertion.)

mechanism for the rectification of errors to ensure fidelity. Currently there is experimental evidence available that discrimination and rectification of errors during protein biosynthesis can occur when a tRNA is loaded with an amino acid (for details see Chapter 3) and at the decoding step.

Biochemical investigations of possible errors seen in translation are very difficult and have not given clearcut results. However, current information suggests that the frequency of mistranslation of mRNA is about once in 3000. This result affirms the great precision of the steps in the biological mechanism which is far greater than that which would be deduced from a study of non-biological chemical reactions. It should be noted that it is necessary to maintain 1 in 3000 accuracy at each discrimination step, not as some biochemists mistakenly propose that this overall accuracy can be achieved by multiplying smaller accuracy factors from each discrimination step.

2.14   Universality of the genetic code

Up to the present time (1983), the general conclusion about the universality of the genetic code still holds for prokaryotes and eukaryotes with minor exceptions, but now we know that mitochondria provide some startling differences. Eukaryotic cells, as we have seen in section 2.10, probably differ from prokaryotic cells in that UGA, a terminator codon in prokaryotic cells, is a codon for Trp in eukaryotic cells.

More differences have been determined in mitochondria, organelles which exist inside eukaryotic cells to provide energy sources through oxidative phosphorylation. Mitochondria possess their own DNA, which contains genes for their specialized enzymes and their rRNAs and tRNAs. Naturally, this DNA varies with the source of mitochondria. Thus in yeast mitochondria, UGA codes for Trp and CUA for Thr. This latter finding was an especial shock, since CUA normally codes for Leu. Furthermore, in human HeLa cell or beef liver mitochondria CUA is normal, but UGA codes for Trp, then AUA and AUU in addition to AUG are codons for initiator Met, and termination is coded for by AGA and AGG (normally Arg codons) as well as UAA and UAG.

The current explanation of these irregularities lies in the decoding properties of appropriate tRNAs. In the case of human mitochondria, the total genomic DNA has been sequenced by F. Sanger's group in Cambridge. Startlingly, only 22 tRNA genes were found rather than the minimum 31 needed by Wobble Theory (section 2.11). Clearly, some tRNA species decode more codons than in bacteria, yeast and animal cells. If the potential of decoding four codon families differing only in the third base by the same tRNA is used then we reach a minimum of 23 tRNAs. Thus each CUN, GUN, UCN, CCN, ACN, GCN, CGN and GGN family would be decoded by one tRNA. The remaining discrepancy of the finding of only 22 tRNAs may be explained by only one tRNA[Met] being found. This presumably must double in some unexplained way as an initiator and an elongator tRNA. To explain the expanded decoding properties of mitochondrial tRNAs, it is possible to extend Wobble Theory to allow U.N and U*.R base pairing. This means that a U in the first position of an anticodon can decode all four bases in the 3rd position of the codon and that a modified U (U*) can decode a purine (A or G). With this proviso for mitochondria so far supported by experimental results, we can still hold on to the universality of the genetic code.

3 Components of Protein Biosynthesis

3.1 Introduction

We can now describe the biochemical details of the transfer of genetic information. The first step in the transfer (see Fig. 3–1 in which the nucleic acids are depicted with a more structural shorthand than used in Chapter 1) is called transcription and is straightforward in the sense that no change in molecular language is concerned, although a large enzyme called RNA polymerase is needed to maintain the fidelity of base sequence in the newly synthesized mRNA. We are more interested here in the second stage of

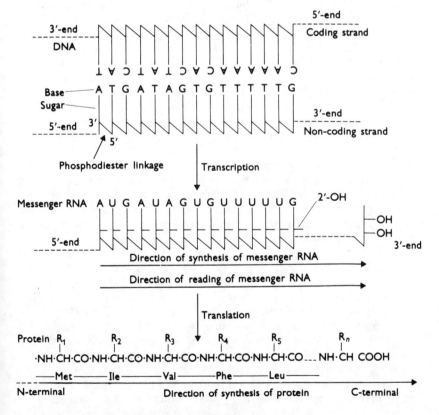

Fig. 3–1 Transmission of genetic information.

information transfer called translation, whereby the information changes languages with more complicated chemical and structural problems. Many components are needed for the translation process and are involved in the biosynthesis of proteins. Most of our current knowledge about translation has arisen from biochemical experiments using various cell-free systems for making proteins.

Some selected properties of the cellular components of protein biosynthesis will be detailed before describing a generally accepted model for the protein elongation step where all the components come together.

3.2   Amino acid activating enzymes

First the amino acid must be activated and carried to the peptide bond forming enzyme. The amino acid is linked catalytically to tRNA, usually termed charging or loading the tRNA, by an amino acid activating enzyme.

In the late 1950s it was found that amino acids could be enzymically bound to a then uncharacterized class of RNA by a soluble protein fraction from rat liver. The protein fraction, which was soluble at pH 7, was isolated by precipitation at pH 5. This so-called pH 5 enzyme fraction was used in the early studies of amino acid activation in protein biosynthesis. The name 'activating enzyme' is used here because it relates its role to the protein biosynthetic reactions under discussion. Alternative names for this enzyme activity are aminoacyl-tRNA synthetase, and the systematic, but cumbersome, amino acid:tRNA ligase (AMP). So far bacterial, bovine pancreas and yeast enzymes have been obtained in a pure state.

Amino acid activating enzymes constitute up to 10% of the cell's protein. This is equivalent to 1000–5000 molecules per cell. (These figures are approximate since the amount may vary under different growth conditions and the molecular weights of many of the enzymes are not known.)

In general we assume that only 20 activating enzymes are found in all biological material. However, in a few cases multiple species of a particular amino acid activating enzyme for the one amino acid are found in eukaryotic cells and may be due to compartmentation where one enzyme is cytoplasmic and one mitochondrial.

Since there is only one activating enzyme for charging several different tRNAs (*isoaccepting species*) with the same amino acid it was hoped that a knowledge of the several tRNA primary structures would reveal how the activating enzyme recognized them. Unfortunately, however, these studies have not been very helpful. When the recognition site of a tRNA for its activating enzyme is discussed an uncertainty arises. There is no guarantee that every tRNA is recognized in a similar fashion by some property of a similarly located set of nucleotides. Perhaps the recognition of particular tRNAs by their specific activating enzymes has evolved differently, so that by now there are different classes of tRNA recognition features. If this is correct no generalization about the recognition process will be possible when details of recognition of just one tRNA by its activating enzyme are known.

Structure

How the activating enzyme distinguishes one tRNA from another is still a mystery in spite of much physical chemical investigation. What is really needed is an X-ray crystallographic study of a complex of an activating enzyme and its specific tRNA. Such a complex has recently been crystallized for the yeast Asp system and there are also now available for X-ray analysis crystals of both tRNAs and activating enzymes. The lysine activating enzyme from yeast was the first to be crystallized, but the crystals were not suitable for high resolution X-ray analysis. At present, good crystals are available for yeast leucine activating enzyme of mol. wt. 120 000 daltons, *Bacillus stearothermophilus* tyrosine activating enzyme (mol. wt. 95 000) and a fragment of *Escherichia coli* methionine activating enzyme of mol. wt. 66 000 containing the part of the enzyme active for loading the tRNA with methionine. The molecular structure of the tyrosine and methionine activating enzymes should soon be known to better than 0.3 nm resolution.

Several activating enzyme species are now available in amounts up to 1 g. The determinations of the sequences of the *E. coli* methionine and *B. stearothermophilus* tyrosine activating enzymes are far advanced, whilst that of tryptophan activating enzyme from *B. stearothermophilus* is completed.

With the recent development of rapid DNA sequencing methods, it has become easier to consider determining the primary structures of these enzymes at the gene rather the protein level. What is required, of course, is a knowledge of the location of the particular genes on a genetic map for isolation of pieces of DNA including the genes. Thus, the *E. coli* genes for Ala, Met, Tyr, Thr and Phe enzymes will soon be sequenced.

Reaction mechanism

Activating enzymes catalyse the esterification of an amino acid to tRNA. It is generally accepted that there are two distinct steps in the mechanism of the charging reaction: (1) the activation reaction and (2) the transfer reaction (Fig. 3–2). An aminoacyladenylate formed in step 1 remains enzyme-bound and transfers its amino acid to tRNA in step 2. Either step can be used to assay the enzyme activity.

The overall charging reaction requires the presence of magnesium ions and satisfies its energy requirements by the splitting of ATP to AMP and inorganic pyrophosphate, PP_i. The optimum pH for different activating enzymes varies. Some work best at a pH between 8 and 9, but the charging of the tRNA is usually carried out at about pH 7.5 because of the instability of the aminoacyl-ester bond formed (Fig. 3–2) even under mildly alkaline conditions.

The aminoacyl-ester bond is of high potential energy, comparable with that of the pyrophosphate bond in ATP. In step 1 of the charging reaction the amino acid is joined to the 5'-phosphate of AMP in an acyl phosphate (a mixed acid anhydride) linkage. This very reactive acyl phosphate, as an enzyme complex (as shown in Fig. 3–2), transfers its acyl group in step 2 to another adenosine residue – the terminal adenosine of the tRNA – in an aminoacyl-ester linkage with a hydroxyl of the terminal 2', 3'-diol group. The point of attachment used

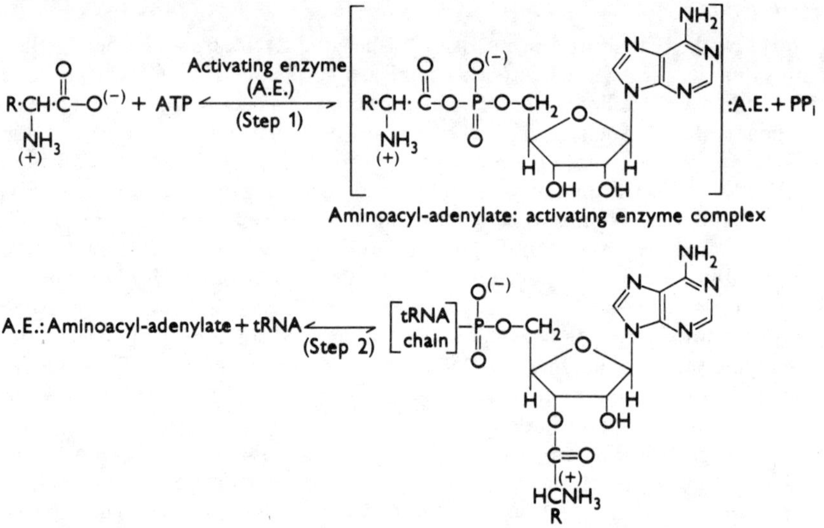

Overall reaction Amino acid + ATP + tRNA $\xrightleftharpoons{A.E.}$ aminoacyl-tRNA + AMP + PP$_1$

Fig. 3–2 Charging the tRNA as a two step process.

to be taken to be the 3′-OH but since aminoacyl migration may occur 10^4 times per second between these two hydroxyl groups – nearly 1000 times faster than peptide bond formation – it is meaningless to distinguish between them for amino acid attachment to tRNA. Actually, evidence is available that some activating enzymes (e.g. phenylalanine) attach the amino acid to the 2′-OH group of the tRNA and that migration to the 3′-OH group is necessary before peptide bond formation can occur. The significance of this migration is unclear. Also, recent studies of the enzyme kinetics have suggested that the reaction mechanism may be more of a concerted one step affair rather than that conveniently described above.

Accuracy in protein biosynthesis

The two steps in the charging reaction clearly permit the activating enzyme to carry out two specific checks for ensuring accuracy in protein biosynthesis.

First, the correct amino acid is recognized by the enzyme in the formation of the aminoacyl-adenylate complex. Amino acid analogues have only a low chance of competing in this reaction. For example, the methionine activating enzyme forms an ethionyl-adenylate complex at one hundredth the rate of formation of the methionyl-adenylate complex.

Even if the activating enzyme were to make a mistake in choosing its amino acid, the second specificity check, for the correct tRNA, is very accurate. Although in general the activating enzymes do not recognize a wrong amino acid and join it to AMP there are exceptions. For example, bacterial isoleucine activating enzyme will form valyl-adenylate albeit with a lower affinity of binding than for the isoleucyl derivative. However, the accuracy of protein synthesis is maintained in the transfer reaction, since the isoleucine enzyme will not form valyl-tRNA from the adenylate complex.

Detection

The activating enzyme is assayed for ability to charge tRNA with an amino acid in a buffered solution containing magnesium (approx. 10 mM); the amino acid, ATP and tRNA are in excess so that the initial rate of esterification of the amino acid to tRNA is dependent upon the amount of enzyme. The aminoacyl-tRNA, radioactively labelled in the amino acid, is precipitated with cold trichloroacetic acid, collected by filtration on cellulose nitrate or cellulose acetate filters and the dried radioactive precipitate is quantitated either by liquid scintillation or gas flow counting. Alternatively, the labelled aminoacyl-tRNA in solution can be spotted on a circle of filter paper which is then washed in cold trichloroacetic acid solution to remove the unreacted labelled substrate. The filter paper is dried in ethanol before its radioactivity is counted in a liquid scintillation counter. (Obviously, the same scheme can be used to assay for the presence of tRNA if the activating enzyme is added in an excess to make the amount of tRNA rate-limiting.)

3.3   The adaptor, tRNA

Function

In the late 1950s it was realized, especially by Crick, that there seemed to be no simple way in which nucleic acids could programme the synthesis of protein by direct structural interactions with amino acids. Thus, Crick proposed his *Adaptor Hypothesis* whereby an adaptor molecule (at that time vaguely specified) would mediate between the amino acid and the nucleic acid which carried information for directing the amino acid sequence. Shortly afterwards, in 1957, Hoagland discovered in a rat extract a type of RNA which could bind amino acids specifically. This discovery of what was soon recognized as the missing adaptor molecule stimulated an ever expanding amount of research into the relationship between structure and function of tRNA.

In addition to carrying esterified and activated amino acids to the ribosomal site for peptide bond formation during protein biosynthesis, tRNAs function by decoding genetic information carried by mRNA.

Synthesis

Although tRNA and rRNA (ribosomal RNA) make up about 20% and 80% of total RNA while the less stable (*in vivo*) mRNA accounts for about 2%, as little as 1% of the total DNA functions as template for the synthesis of both tRNA and rRNA, in roughly equal amounts. Evidence for the genome content of tRNA and rRNA comes from hybridization studies. (Hybridization involves the formation of stable double-strands, containing single strands of a DNA gene with its specific complementary RNA.)

The total is estimated to make about 1% of the bacterial cell's dry weight. In a rapidly growing bacterial cell there are of the order of 4×10^5 tRNA molecules of perhaps 50 different types (the exact number of species is not known). There is no definite estimate of how much of the tRNA in the cell is charged with an amino acid, but it is likely that the catalytic activity of the activating eyzmes is capable of keeping the cellular tRNAs fully charged for peptide bond formation if there are sufficient free amino acids available.

A transfer RNA is made in the cell by DNA-dependent RNA polymerase which copies a length of DNA (a gene) specifying a particular tRNA. An *Escherichia coli* chromosome contains different genes for the 50 or so different tRNA molecules but multiple copies of each gene may exist and the number of copies may vary for different tRNAs. About 0.5% of the chromosome appears to code for tRNA genes.

Isolation from cells

Transfer RNA is usually extracted from bacterial cells with buffered aqueous phenol. Addition of ethyl alcohol to the aqueous layer precipitates the tRNA; the phenol layer contains proteins and cell debris. Several further precipitations from salt solution, then an extraction of the tRNA into methoxyethanol from phosphate buffer to remove polysaccharides and traces of DNA, and a final ethanol precipitation from an aqueous organic mixture yields a tRNA preparation which can be used for functional tests or further purified by column chromatography to give the individual tRNA species.

Role of tRNA in decoding

The direct participation of tRNA as a carrier of the amino acid in protein synthesis was first shown in studies using a bacterial cell-free system (see Chapter 2).

Then in 1962 a group of biochemists proved that the interaction of the aminoacyl-tRNA with a specific coding sequence in messenger RNA is independent of the amino acid in the complex. Thus, once the amino acid is attached to the tRNA by its specific activating enzyme it plays no part in messenger recognition; the information is carried by the tRNA adaptor. In a classical experiment CHAPEVILLE *et al.* (1962) altered the cysteine residue of charged Cys-tRNA^{Cys} to alanine by hydrogenation over Raney nickel, giving Ala-tRNA^{Cys}. From earlier studies using the cell-free system it was known that poly (U, G) random copolymer stimulated the incorporation of cysteine but not of alanine. However, when the synthetic Ala-tRNA^{Cys} was added to the cell-free system programmed by poly (U, G) then Ala was incorporated into the polypeptide. Thus, the alteration of the amino acid did not alter the coding response of the tRNA to which the amino acid was attached. This was confirmed later in a mammalian system.

Purification of tRNA species

Pure tRNAs were required for two reasons: to provide pure aminoacyl-tRNAs for work on the genetic code (as described in Chapter 2) and to enable determination of their nucleotide sequences (primary structures).

Since all tRNA molecules have similar properties and are all about 80 nucleotides long, a combination of fractionation methods is usually required to achieve complete purification of a single species. Countercurrent distribution was notable among the early methods used. It is still used occasionally but has been replaced mainly by different types of column chromatography.

The most widely applicable column chromatographic methods involve respectively (1) an inert diatomaceous earth called chromosorb (a reversed phase method); (2) Diethylaminoethyl-Sephadex (DEAE-Sephadex;) (3) benzoylated DEAE-cellulose. Using a combination of these methods hundreds of milligrams of pure species of tRNA have been isolated for determination of primary structure and for crystallization attempts. For more rapid sequence determination of tRNAs some of the chromatographic methods, suitably scaled down, can be used to purify the 0.5 mg of ^{32}P-radioactively labelled tRNA needed for this purpose.

Furthermore, the recent development of rapid RNA sequencing methods using electrophoresis on polyacrylamide gels where the tRNA fragments are radioactively labelled after they have been prepared, allows the determination of a tRNA sequence on as little as 10 μg of material. This amount can be purified by thin layer chromatography and polyacrylamide gel eletrophoresis, which is more convenient than the older conventional methods above.

The most reliable test of purity of tRNA is proof of a unique nucleotide sequence, but sometimes this is not feasible. Unlabelled tRNA is generally estimated by means of a charging test. If the molar ratio of the amino acid to tRNA approaches unity when the tRNA is enzymically charged under optimal conditions, the tRNA is taken to be pure. Of course, this assumes that the sample does not contain isoaccepting species. This possibility is checked by using several different separation systems.

Primary structure

Methods of determining the primary structure of tRNAs are outside the scope of this book (for more information see BARRELL, 1971). The task of sequencing tRNA is made easier by the presence of unusual or rare bases which can be used as reference points. (For examples of rare bases see BARRELL and CLARK, 1974; SPRINZL and GAUSS, 1982.) Many of the rare bases are normal bases methylated at specific positions by enzymes called methylases during tRNA biosynthesis.

The sedimentation coefficient of tRNA (about 80 nucleotides long and mol. wt. about 25 000) determined by sucrose density gradient centrifugation is 4 S. At least 230 different tRNA sequences are known. All tRNAs start with a free phosphate at the 5'-end of the polynucleotide and finish with a common sequence CpCpA at the 3'-end. The 3'-adenosine has a free 2', 3'-diol to which the amino acid becomes attached as already described. The remarkable feature of the known primary structures is that they can all be fitted to a base-paired secondary structure – the 'clover leaf' structure (Fig. 3–3) – first proposed by Holley for yeast alanine tRNA (HOLLEY *et al.*, 1965). Such a clover-leaf structure will accommodate the constant features of most known tRNA sequences (Fig. 3–3). The exceptions so far include (1) yeast and *Escherichia coli* initiator methionine tRNAs, (2) tRNAs involved in cell wall metabolism rather than protein synthesis, (3) a *Salmonella typhimurium* histidine tRNA and (4) mitochondrial tRNAs.

Most regions of the tRNA structure are remarkably constant, for example stems a, c, and e have 7, 5 and 5 base pairs, respectively, and loops II and IV each contain 7 non-base-paired nucleotides. The variable regions are confined to stems b and d and loops I and III.

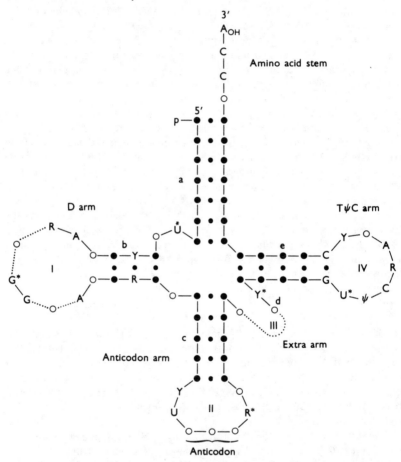

Fig. 3–3 Generalized structure for a tRNA in a clover-leaf form illustrating the potential secondary structure. Full circles are H-bonded bases in base pairs. Open circles are bases not in clover-leaf base pairs. H-bonds in base pairs are represented by big dots. R = purine base; Y = pyrimidine base; Ψ = rare base pseudouridine. * indicates that the nucleotide may be modified. Base-paired regions (stems) are lettered a to e and non-base-paired bases are in loops I to IV. The dotted parts of loops I and III indicate variation in number of nucleotides. Also the arms of the clover leaf have trivial names as shown: an arm = a stem + a loop.

Actually, normal tRNAs can be subclassified as small or large according to the size of the extra arm. When this is just a loop, as in most cases, the length range is in the region 75 to 80 giving the small subclass. Large tRNAs contain an extra arm giving a length range of 85 to 95. Large tRNAs are confined to isoaccepting species for Tyr, Leu and Ser.

There is much physical chemical evidence supporting the clover leaf arrangement in two dimensions. Physicochemical studies measuring ultraviolet absorption changes with increasing temperature (melting curves), nuclear magnetic resonance studies identifying base-paired protons, Raman and infrared spectral studies and tritium exchange studies have all given evidence for base-paired helical regions in tRNA. Light-scattering measurements and low angle X-ray scattering in solution have suggested that the tRNA molecule is long and thin. Many other experiments involving susceptibility of parts of the structure to enzymic cleavage, or to chemical reagents thought to be specific for single-stranded regions have confirmed the availability of the putative bases of loops I and II for modification. Chemical modification has been used to investigate the folding of the tRNA molecules but suffers from the drawback that reaction at one site may cause partial unfolding and give misleading results. However, one fact that seems certain from experiments of this type is that loop IV is somehow buried in the tertiary structure so that it is not available for reaction unless the molecule is unfolded by heating. The only unambiguous way of obtaining tertiary structural details is X-ray analysis. Since 1968 (CLARK et al., 1968) many different tRNA species have been crystallized. However, only crystals of yeast tRNA^{Phe} and yeast tRNA^{Asp} have so far yielded high resolution X-ray diffraction patterns. Because the X-ray work progressed slowly, there have been ingenious attempts at building models for the tertiary structure of tRNA based on the available biochemical and physicochemical evidence, but none of them predicted the correct shape of the molecule. This has been determined to 0.3 nm resolution for yeast tRNA^{Phe} by X-ray analysis by two groups of workers (ROBERTUS et al., 1974; KIM et al., 1974) (Fig. 3–4) and later extended to 0.25 nm resolution by the same workers. This was the first nucleic acid three dimensional structure to be determined by X-ray crystallography. It is likely that the structure contains general features for all tRNA structures. To obtain the structure shown in Fig. 3–4 from the two-dimensional arrangement in Fig. 3–3, the amino acid stem is stacked on top of the TΨC-stem to form a long helical bar in an overall shape which suggests a T structure (also called L by other workers, RICH and KIM, 1978). The other stems, the D-stem and anticodon stem are also stacked approximately on each other to give the other bar of the T. However, the central region contains many crossing tertiary triple base H-bonding interactions to hold the structure in the form of the T so that the D-stem is augmented in length and width. Clearly, when functioning, the decoding region of the tRNA (the anticodon) is exposed for interacting with mRNA and is at the opposite end of the molecule to where the amino acid is attached. At present we do not know with any certainty how the amino acid activating enzyme recognizes the tRNA.

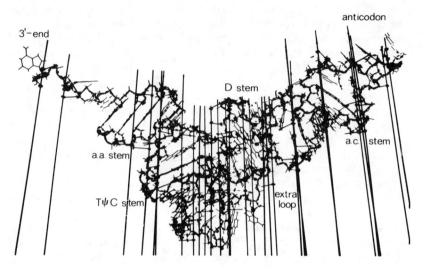

Fig. 3–4 Molecular model of yeast phenylalanine tRNA at 0.3 nm resolution shown in silhouette form. (Courtesy of MRC Laboratory of Molecular Biology, Cambridge.)

Very recently, a high resolution structure has been determined for yeast tRNA^{Asp} and less well defined structures are now known for yeast and *E. coli* tRNA$_f^{Met}$ species. All of these structures do not differ significantly in overall shape from that of yeast tRNA^{Phe}.

The next step in protein synthesis after the tRNA is charged brings two aminoacyl-tRNAs together to allow a peptide bond to be formed.

3.4 Ribosomes

Peptide bond formation takes place on a large ribonucleoprotein workbench in the cell called the ribosome (Figs 3–6, 3–7).

Once again most of the detailed information about ribosomes comes from studies using bacterial cell-free systems, although it has recently been possible to isolate very active ribosomes from extracts of mammalian cells, such as reticulocytes, liver and ascites tumour cells, and plant cells such as wheat germ.

Protein synthesis never appears to occur free in solution but only on ribosomal surfaces. The ribosomal structure is designed to orient two charged tRNA molecules specified by mRNA, in such a way as to permit the formation of a peptide bond between a polypeptide and an incoming amino acid.

There is good evidence that there are two ribosomal sites for the aminoacyl-tRNA when a peptide bond is being formed (see Fig. 3–7). So far we know very little about the detailed structures of these sites and the evidence for them is rather circumstantial. The sites position the peptidyl-tRNA (P-site for tRNA$_n$ in Fig. 3–7A) and aminoacyl-tRNA (A-site for tRNA$_{n+1}$ in Fig. 3–7A) during the propagation or *elongation* phase of protein biosynthesis. The concept of two sites may need some revision for special stages of protein synthesis such as starting, but this is not completely clear at present. The bacterial ribosomal unit is called a 70 S particle because of its sedimentation properties in the ultracentrifuge.

Both prokaryotic and eukarytic cells contain ribosomes composed of two subunits, one of which is about double the size of the other. In bacteria the ribosomes and their subunits have sedimentation constants of 70 S, and 50 S and 30 S repectively. In each case the ratio of the amount in mass units of RNA to protein is 63:37. The ribosomes found in eukaryotic cytoplasm seem to have a higher protein content and larger RNA molecules, and have sedimentation coefficients of 80 S, 60 S and 40 S respectively for the single ribosome and its subunits. Ribosomes in eukaryotic cells appear to be rigidly ordered in a cellular membrane called the endoplasmic reticulum and this makes their isolation with functional integrity more difficult. Indeed ribosomes from mammalian cells when isolated often contain extraneous membrane material. An interesting feature of eukaryotic cells is the occurrence in organelles such as mitochondria and chloroplasts of smaller ribosomes, somewhat similar in properties to those of bacteria.

In rapidly growing (log phase) bacterial cells (generation time of about 20 min) there are of the order of $(15-18) \times 10^3$ ribosomes. Each ribosome has a molecular weight of about 2.7×10^6 daltons and has dimensions about $15 \times 15 \times 20$ nm. Since an average bacterial cell 2 μm long and 1 μm across has a wet weight of about 10^{12} daltons, it can be estimated that in total the ribosomes make up nearly a quarter of the total cellular mass (i.e. the cell's dry weight). Thus a good proportion of the cell is devoted to the business of making proteins. Only one protein is made at a time on a ribosome. When the protein is finished the ribosome moves to the next starting point on a mRNA which codes for more than one protein or is released and can take part in the synthesis of another type of protein as programmed by a newly attached mRNA. Before the attachment of new mRNA the ribosome is dissociated into subunits.

Polyribosomes and subunits

Depending on the ionic environment, ribosomes can exist in a variety of associated or dissociated states which may or may not be related to functional states within the cell. In the native state one 50 S particle and one 30 S particle combine giving a single 70 S particle. Higher sedimenting states found in the cell are due to several monomeric 70 S particles attached to a strand of mRNA. Indeed, if a cell is very carefully broken it is possible to isolate most of the ribosomes in polyribosomal formation with several ribosomes (all making the same proteins) strung along a single-stranded mRNA like beads on a string. It has been established, in both bacterial and mammalian cells, that the functional formations for protein synthesis are polyribosomes, for example for the synthesis of globin in reticulocytes they are quite often of the order of 200 S, containing five ribosomes on a single mRNA. These polyribosome formations have been confirmed by electron microscopy. Indeed, the elegant studies of Miller and his colleagues (e.g. MILLER, 1973) (Fig. 3–5) show that coupling of transcription and translation often occurs in the cells; polyribosomes showing

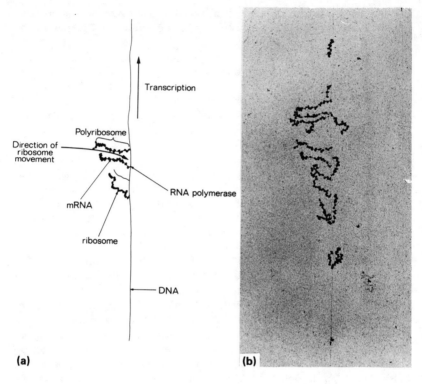

Transcription

Polyribosome

Direction of
ribosome
movement

mRNA

ribosome

RNA polymerase

DNA

(a)   **(b)**

Fig. 3–5 Diagram (**a**) explains the electron photomicrograph (**b**) showing genes in action. Magnification is approximately × 27 000. (Courtesy of Drs B. A. Hamkalo and O. L. Miller, Biology Division, Oak Ridge National Laboratory, U.S.A.)

translational activity can be seen to be attached to new mRNA being transcribed off DNA.

Very recently, electron microscopic visualization of genes in action has been developed so that even nascent polypeptides can be seen growing on ribosomes where the mRNA is being translated (FRANKE *et al.*, 1982). These beautiful examples of the visualization of active genes clearly place the interpretation of biochemical studies on a solid realistic footing for the explanation of what is happening in cellular protein biosynthesis.

During the synthesis of proteins of mol. wt. 30 000–50 000 daltons (300–500 amino acids) about 12 to 20 ribosomes will be attached to the mRNA molecules. At maximal utilization of mRNA length, taking the diameter of the ribosome to be 20 nm, there could be one ribosome per 60 nucleotides on the mRNA. In this context, when the mRNA attached to a ribosome is digested with a ribonuclease, the ribosome is found to protect a stretch of about 30 nucleotides from enzymic degradation. This must be a measure of the amount of mRNA which is not exposed.

The ability of a single mRNA to function simultaneously on several ribosomes helps to explain why a cell needs so little mRNA (only 1–2% of total RNA). Polyribosomes are clearly an efficient and economical way of using mRNAs.

Preparation of 70 S ribosomes has been described in Chapter 2. When bacterial cells are broken by grinding with acid-washed alumina in the absence of inhibitors, the polyribosomes are nearly all degraded to 70 S ribosomes because released ribonucleases degrade the mRNA.

To what extent 70 S ribosomes exist free in the cell is debatable. There is reasonable evidence for the existence of about 10–20% of the ribosomes in the 70 S form, whereas about 70% are incorporated into polyribosomes. The remaining 10–20% (but this appears to vary with the bacterial strain) exist in their subunit form.

It is known that 50 S and 30 S subunits recycle through a round of protein synthesis, i.e. after one protein chain has been synthesized using a 70 S particle the subunits can dissociate. This agrees with the idea that the 30 S subunit joins alone to mRNA before the 50 S unit when synthesis of polypeptide chains is started. Actually, in the case of bacteria concerning polycistronic mRNA (see section 3.5), the 70 S particle probably does not dissociate until the end of the multigene message.

The most likely explanation for the existence of two subunits in the functional ribosome for peptide bond formation is that two subunits are necessary for movement along the mRNA. Indeed it is possible to construct models whereby an oscillation between the subunits causes the mRNA to move over the ribosome. Perhaps the ribosome is not of such a passive character as is suggested by referring to it as the site for protein biosynthesis.

Components

The ribosomal RNA is an integral part of ribosomal structure unlike mRNA because when rRNA is removed the ribosomal structure, and hence the functional properties, are destroyed. When ribosomes are extracted with a mixture of phenol and water in the presence of a ribonuclease inhibitor, the RNA is usually extracted into the aqueous phase. Bacterial ribosomes extracted carefully in this way release equimolar amounts of 23 S, 16 S and 5 S RNA components. If the subunits are separated before their RNA is extracted, it is found that both the 23 S and 5 S come from the 50 S subunit, whereas only the 16 S RNA comes from the 30 S subunit (Fig. 3–6).

The molecular weights of the bacterial rRNA species are about 10^6, 5.5×10^5 and 4.0×10^4 daltons for 23 S, 16 S and 5 S respectively. Hybridization experiments show that about 0.4% of the bacterial chromosome codes for rRNA. Thus, there are about six genes for each of 23 S, 16 S and 5 S rRNAs. *Escherichia coli* rRNA genes are closely linked and some have been shown to be adjacent, with 5 S, 23 S and 16 S in tandem formation. Although it is likely that the precursor rRNA could, therefore, contain the three rRNAs with perhaps spacer sequences including tRNA genes linking them, there is also electron microscopic evidence for a least the 16 S and 23 S rRNAs being

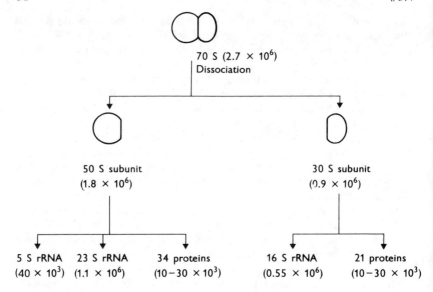

Fig. 3–6 Components of a 70 S ribosome. Molecular weights of components in daltons are shown in brackets.

transcribed in distinct pieces off the adjacent genes. Perhaps different possibilities occur for different gene sets. This contrasts with the situation in mammalian cells where rRNA genes can occur in hundreds of copies and are transcribed as one unit in a special part of the nucleus, the nucleolus, to give a large 45 S RNA. The 45 S RNA from *Xenopus laevis* or human HeLa cells is degraded in a series of steps yielding the final 18 S and 28 S rRNA for the 40 S and 60 S subunits, respectively. The 5 S rRNA appears to be transcribed separately. There is another small RNA, a 5.8 S rRNA, which is also part of the mammalian 60 S subunit and is found in the 45 S RNA precursor.

The rRNAs are largely single strands but appear to contain double-stranded regions as a result of base-pairing. The 23 S and 16 S RNA contain small amounts of methylated bases although the proportion of methylated bases is much smaller than for tRNA. The 5 S RNA, which is only about 120 nucleotides long, does not appear to be covalently bound to the 23 S RNA in the 50 S particle, nor does it contain methylated bases. This raises the possibility that it has a different function from the longer RNAs, which have structural importance. Some unsubstantiated proposals have been made, for example that 5 S RNA is involved in polypeptide chain termination or is a structural part of the A-site for binding aminoacyl-tRNA.

The total sequences of bacterial 5 S, 16 S and 23 S rRNAs are now known. The larger rRNAs usually now have their sequences determined at the gene level by rapid DNA sequencing. Currently, several large eukaryotic rRNA and their precursors are being sequenced.

As indicated in Fig. 3–6, the protein components of the ribosome have been separated and most are well characterized, enabling a start to be made by several research laboratories on the immense job of determining the structure of the ribosome. Indeed, the primary structures of all of the ribosomal proteins from *E. coli* have been elucidated thanks mainly to the laboratory of H. G. Wittmann in Berlin. Furthermore, a few ribosomal proteins have been crystallized for X-ray diffraction studies. Our knowledge of the ribosomal components in terms of their function and role in the ribosome's structure is still very sketchy and preliminary. Some significant information on the protein packing has come from spectacular experiments which were successful in reconstituting active bacterial subunits from their constituent protein and nucleic acid parts. This was also another pertinent example of automatic self-assembly into correct supramolecular particles without external genetic information. In this field, other methods such as immunoelectron microscopy and neutron diffraction have also enabled steady progress recently in locating the relative positions of surface and embedded ribosomal proteins. One should also mention the current idea that rRNA is a scaffold for the proteins so its conformation is not radically changed on binding them.

The final component of protein biosynthesis we need to discuss is the programming element for ordering the amino acids in the protein's primary structure. As we have seen Chapter 2, it is messenger RNA (mRNA).

3.5   mRNA

Properties

Messenger RNA is the intermediary in the transfer of information from DNA to the protein-synthesizing sites on the ribosome. *The sequence of nucleotides in mRNA directly programmes the order of amino acids in a protein synthesized on a ribosome.* The elucidation of the genetic code revealed that mRNA is translated into protein by a process in which a sequence of three nucleotides (a triplet) signifies one amino acid (see Chapter 2).

Although mRNA comprises only about 3% of total cellular RNA, it was thought until recently that in bacteria up to 99% of the total DNA (genome) is transcribed into mRNA. This apparent contradiction is resolved because the function of mRNA involves greater molar quantities of rRNA and tRNA and although tRNA and rRNA are quite stable and have low turnover rates, in general bacterial mRNA is relatively unstable. In contrast, mRNA of eukaryotic cells is thought to be much more stable. For example, the half-life of the mRNA in rat liver is about 5 h at 30°C whereas that of bacterial mRNA is a few minutes. Recent research on chromosome structure has revealed that varying amounts, perhaps only 10% of the eukaryotic DNA codes for mRNA although it is still likely that 50% or more of the bacterial genome is transcribed into mRNA.

There are about 1000 mRNA molecules in the bacterial cell but they vary greatly in chain length, reflecting the different proteins needed by the cell. The average *Escherichia coli* polypeptide chain contains 300–500 amino acids so that the average size of the mRNA is 900–1500 nucleotides. However, some mRNAs carry information for more than one polypeptide chain. Because of this, they are called *polycistronic* mRNAs and clearly contain at least enough nucleotides to signify the appropriate proteins. They may also contain some sequences of nucleotides used as signals for starting and stopping chains or for defining ribosomal attachment points. Usually polycistronic messengers are translated into polypeptides which have related functions. For example, ten enzymes in the pathway needed to synthesize the amino acid histidine are programmed by a polycistronic mRNA containing about 12 000 nucleotides, i.e. having a molecular weight of about 4×10^6 daltons.

Eukaryotic mRNAs usually have modifications at both their 5'- and 3'-ends. The 5'-ends have the modified nucleoside 7-methylguanosine (m^7G) joined to the mRNA in an unusual 5'-5' pyrophosphate linkage. This distinctive structure has been called a *cap* structure for the 5'-end. In addition, the 2'-OH groups of the sugars of the first two nucleotides on the mRNA chain may or may not be methylated. At the 3'-ends, most eukaryotic mRNAs have a poly A tail some 150 to 200 residues long. The poly A tail is not essential for translation ability *in vitro*, but the cap structure seems to enhance the translation efficiency.

The ultimate proof of a mRNA's character is the determination of its nucleotide sequence, permitting a clear correlation with a known protein's amino acid sequence. It is not safe to say that the technical difficulties associated with this structural test mean that only functional tests will be feasible for some time because of recent progress in developing rapid sequencing methods.

Synthesis

Although double-stranded DNA acts as template for the enzyme RNA polymerase which makes all types of RNA, only one strand of the DNA appears to be transcribed for a given gene. The DNA strand which is copied in complementary and antiparallel fashion is known as the *coding* strand (Fig. 3–1). The role of the other DNA strand is not known but in sequencing studies, since it has the same sequence as the mRNA, it is called the *sense* strand. It is also unclear whether the RNA polymerase makes one strand of polycistronic mRNA by reading double-stranded DNA, or whether the enzyme separates the DNA strands before copying one of them. The direction of synthesis is from the 5'-phosphate end to the 3'-diol end. The mRNA is also read (or decoded) in groups of three nucleotides (triplets) from the 5'-end to 3'-end from the special starting point which has not yet been completely defined. It is likely that nascent mRNA is protected from degradation by addition of ribosomes; if a ribosome fails to bind to the beginning of the mRNA, degradation could begin. Degradation of a long polycistronic mRNA probably starts before translation is complete – an added hindrance to the biochemist aiming to isolate an intact mRNA. Coupling of translation and transcription can clearly occur as evidenced by the extraordinary electron photomicrograph in Fig. 3–5. Thus the mRNA is translatable before it is completely synthesized.

In eukaryotic cells, primary transcripts are not used directly as mRNA. Rather the long transcripts are processed extensively before being transported from the nucleus to the cytoplasm. Thus translation and transcription are spatially and temporally separated in eukaryotes, whereas they are coupled in prokaryotes. The eukaryotic primary transcripts can vary in length from 2000 to 20 000 nucleotides giving a very mixed population of RNA molecules called heterogeneous nuclear RNA (hnRNA). The hnRNA molecules are much longer than mRNAs which are derived from them by cleavage and relinking (called splicing) by special enzymes (see Fig. 2–6). It is possible that hnRNA has other functional roles as well. In addition, the eukaryotic mRNAs must be modified to add a *cap* at their 5′-ends and usually a long tail of poly A at their 3′-ends. Again the definitive roles of these modifications are unknown, but perhaps protection against degradation or signals for protein binding during transport are good candidates for protective roles.

mRNA in cell-free systems

Bacteriophage mRNA stands out as being the most readily available natural messenger for studies on protein biosynthesis. A bacterial cell-free system supplied with RNA extracted from bacteriophage containing single-stranded RNA will synthesize protein which is characteristic of the bacteriophage. This natural mRNA can be extracted from the bacteriophage in such a way that its integrity with respect to starting signals, ribosomal attachment points and secondary structure, is maintained. Although single-stranded RNA from plant viruses, such as tobacco mosaic virus, will make polypeptides in a bacterial cell-free system, the products have not been definitively characterized as viral proteins; translation of these viral RNAs appears to be more correct, as expected, in extracts of wheat germ or rabbit reticulocytes. A fair test of a plant virus RNA's ability to act as messenger has been hampered until recently by the lack of very active plant cell extracts. In contrast, it has been possible to assemble mammalian cell extracts which are very active in synthesizing proteins. Single-stranded animal virus RNAs will work as mRNAs in mammalian cell extracts, for example encephalomyocarditis virus RNA can be translated actively in an ascites tumour cell extract. Although animal viral messengers are not as well characterized as bacteriophage messengers, many mammalian mRNAs have now been isolated and translated *in vitro*. The list of isolated and tested mammalian mRNAs is rapidly being extended. In addition to globin mRNA, eye lens α-crystallin mRNA, myosin mRNA, ovalbumin mRNA and myeloma protein mRNA are amongst the more useful to have been identified. It is of interest that all eukaryotic mRNAs so far identified are monocistronic. If this proves to be a general property of eukaryotic mRNAs (non-viral in origin) it will contrast sharply with the bacterial situation.

3.6   Cyclic scheme for protein biosynthesis

Peptide bond formation

Although a peptide bond might be expected to form spontaneously between two aminoacyl-tRNAs in the correct ribosomal sites it has been established that the formation is brought about enzymically; the enzyme, *peptidyl transferase*, is

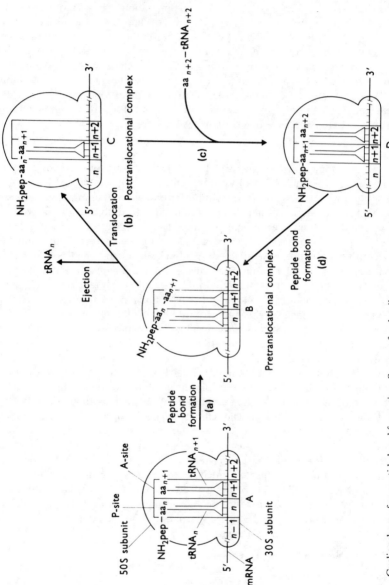

Fig. 3–7 Cyclic scheme for peptide bond formation. See text for details.

an integral part of the ribosome. Supernatant protein factors called elongation factors have also been implicated in the elongation phase of biosynthesis. At present we believe that aminoacyl-tRNA is carried to the ribosomal A-site in a ternary complex with an elongation factor and a cofactor, guanosine triphosphate.

Elongation

After polypeptide chain initiation a series of peptide chain elongation steps occurs before the process of chain termination releases the completed protein. Elongation is defined as the addition of amino acids one at a time to a growing polypeptide in a sequence programmed by mRNA.

The generally accepted scheme for peptide chain elongation which involves relative movement of the mRNA and ribosome is shown as a cyclic scheme for peptide bond formation in Fig. 3–7. This scheme is simplified in that various superficial factors interacting with the tRNA or ribosome are omitted and that a static view of the ribosome involving special tRNA binding sites is assumed. At present there is not enough evidence to suggest that we drop the latter view and think in terms of a more dynamic situation involving activated binding states.

The cyclic scheme shown is really self-explanatory, starting with a situation in state A with a growing peptide attached to $tRNA_n$ in the peptidyl-tRNA binding site (P-site) of a 70 S ribosome decoding codon n of the mRNA, and an aminoacyl-$tRNA_{n+1}$ decoding codon $n + 1$ in the aminoacyl-tRNA binding site (A-site). The mRNA is bound to the 30 S subunit and the tRNA stretches across both 30 S and 50 S subunits. The peptide bond is made by the enzyme peptidyl transferase on the 50 S subunit in step (a) leaving, in state B, an uncharged $tRNA_n$ in the P-site and a new peptide extended by one amino acid, aa_{n+1}, attached to $tRNA_{n+1}$ in the A-site. Movement of the $tRNA_{n+1}$ and mRNA now occurs in step (b) to free the A-site state (C) for a new incoming aminoacyl-$tRNA_{n+2}$ in step (c). Step (b), involving movement of $tRNA_{n+1}$ with concomitant ejection of $tRNA_n$, is usually called translocation. When the new aminoacyl-$tRNA_{n+2}$ is bound in the A-site as in state D, the ribosome is back to a state equivalent to A ready for a new round of peptide bond formation and translocation giving the cyclic feature to the scheme. The molecular details and driving forces for the process of translocation and ejection of tRNA are still not elucidated.

4 Punctuation

4.1 Initiation step of protein synthesis

The mechanism by which protein synthesis begins and ends both in bacterial and in eukaryotic cells is now well understood. The signalling mechanism for the control of these protein biosynthetic steps is usually termed punctuation.

Proteins start in a slightly different way in mammalian and plant cells compared with bacterial cells – at least as far as cytoplasmic synthesis is concerned; there is, however, evidence that in cytoplasmic organelles, such as mitochondria, proteins may start by a similar mechanism as in bacteria. Polypeptide chain initiation has attracted much interest because it is generally thought that this would be the most obvious point of control of protein biosynthesis.

Synthetic polyribonucleotides containing a variety of different bases were shown to behave as artificial mRNA in bacterial cell-free systems with reasonable and comparable efficiencies; if a starting signal existed it would be expected that only the polyribonucleotide containing the starting signal would lead to the formation of polypeptides. In addition, cell-free systems programmed with a synthetic messenger (e.g. poly A) gave polypeptide products (e.g. oligolysines) containing no modified end units. Thus, for some time during the early period in the elucidation of the genetic code it was assumed that no special starting signal was needed in the mRNA and that no special protein factor was involved in starting the polypeptide chain. The assumption was made that any tRNA specified by the first codon in a messenger RNA would be able to initiate protein synthesis. This simple view of the mechanism of polypeptide chain initiation arose as an artefact of the cell-free system which allows chains to start erroneously under special conditions.

The currently accepted mechanism of polypeptide chain initiation in bacteria arises from the discovery by MARCKER and SANGER (1964) of the formation of N-formylmethionyl-tRNA *in vivo* and *in vitro*. The terminology indicates that the formyl group (HCO—) is attached to the amino group of the methionine residue. In N-formylmethionyl-tRNA the amino group is thus blocked and hence no longer available for formation of a peptide bond; N-formylmethionine is thus restricted to the N-terminal position in the polypeptide chain. There was the immediate suspicion that this particular tRNA was somehow involved in chain initiation. We were able to show that this hypothesis was correct, using a bacterial cell-free system (CLARK and MARCKER, 1966) (Table 1–1).

The discovery of the initiator tRNA came from a study of the esterification of methionine to its specific tRNA in *Escherichia coli*. Radioactive methionine (in this case labelled with ^{35}S) was incorporated into Met-tRNA with a crude *E. coli* extract containing the methionine activating enzyme. Two radioactive products resulted from digestion of the charged tRNA by pancreatic ribonuclease; one was the expected methionyl-adenosine and the other was an unexpected product, identified chemically as formylmethionyl-adenosine. The latter could also be detected in charged tRNA isolated from growing cells, showing that its formation in the cell-free system is not an artefact.

The maximum amount of formylation of methionyl-tRNA was determined to be about 60–70% of the total methionine accepting activity. This finding suggested that there were two species of methionyl-tRNA in *E. coli* in a ratio of about 70:30, and that only the more abundant species could be formylated. Direct confirmation of this suggestion came when we separated two classes of methionine-accepting tRNAs from each other.

During initiation of protein biosynthesis the mRNA is attached to the 30 S subunit, probably under the influence of protein factors called initiation factors and the initiator tRNA. The exact location on the 30 S subunit is unknown. For translation to proceed, the 50 S subunit must be joined to the 30 S subunit. In the case of prokaryotes, there are at least 3 types of initiation factors and the cofactor GTP known to be concerned in the initiation step, but further discussion of their complicated and still poorly defined molecular interactions is outside the scope of this book. The situation is even more complicated in eukaryotes, where of the order of 10 protein factors have been considered to be initiation factors.

There is reasonably suggestive evidence that the special initiator tRNA binds to a site on the prokaryotic 30 S subunit which becomes, in conjunction with the contribution of the 50 S subunit, the P-site of the 70 S ribosome. A simple view of the initiation step on the 70 S ribosome is illustrated by Fig. 4–1. In this scheme, initiation is shown to occur at the 5'-end of the mRNA, whereas it is now believed that initiation starts internally in mRNAs at special signals involving the codon AUG. The piece of mRNA sequence is decoded by the appropriate tRNA species in the figure. The $tRNA_f$ represents the initiator tRNA and $tRNA_m$ a normal methionine tRNA. Possibly, in the situation concerning initiation points at cistronic beginnings on polycistronic mRNA other than the start of the first cistron, the 70 S ribosome stays on the mRNA between cistrons so that the initiator RNA binds in this case to 70 S not to a 30 S ribosome subunit.

As has been previously stated, eukaryotic mRNAs are monocistronic so that we consider the 40 S ribosomal subunit binds to the mRNA in the initiation process. A strange but unexplained difference between prokaryotic and eukaryotic initiation concerns the order of binding of the initiator tRNA to the ribosome. The special bacterial initiator tRNA ($fMet-tRNA_f$) is mRNA coded for binding on the 30 S or 70 S ribosome, but the eukaryotic cytoplasmic initiator tRNA $Met-tRNA_f$ (also a special methionine tRNA but not formylated) is bound to the 40 S ribosome subunit before the mRNA is attached.

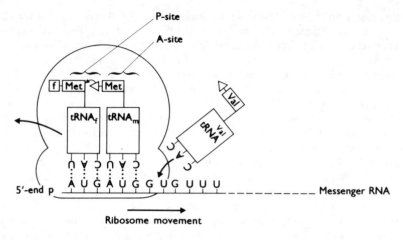

Fig. 4–1   Simple scheme illustrating the initiation step on a ribosome.

4.2   The initiation signal

The decoding properties of the initiator tRNA and the nucleotide sequences around the initiating codon of many different cistrons are now known, so that a reasonable generalization can be made about the nature of the starting signal for protein synthesis. *In vitro* studies indicate that the bacterial initiator tRNA is coded by both AUG and GUG. Until recently, only AUG had been found in the known mRNA cistron starting sequence. GUG has now been found at the start of cistrons in a bacteriophage RNA and *E. coli*. Intriguingly, UUG has also been found to start a bacterial gene and this codon was only weakly active *in vitro*. Once again, mammalian mitochondria are unusual so that for their mRNAs, AUA and AUU in addition to AUG may be starting signals.

There is good evidence that AUG alone is not enough for the starting signal so that the signal could well include, or have close to it, a sequence or structure responsible for selecting the correct initiation codon and to bind the 30 S subunit to the mRNA. The first sequences determined for the initiation sites on mRNA were mostly for bacteriophage RNA where it was possible to correlate determined nucleotide sequences with known amino acid sequences of the bacteriophage proteins. Small pieces of RNA suitable for sequence determination were isolated from experiments in which 70 S ribosomal/mRNA initiation complexes were digested with pancreatic ribonuclease; the initiator tRNA located the ribosome at the initiation site on the mRNA, and the ribosome protected the mRNA against digestion around the appropriate initiation signal. A few examples of the determined sequences (STEITZ, 1969) are shown in Fig. 4–2. The coat-protein and A-protein cistron starting sequence can be arranged in secondary structure to give arms as in the tRNA clover leaf, but the replicase starting sequence cannot. It is possible that an initiation factor has something to do with recognizing and unfolding such sites containing secondary structure.

Beginning of mRNA for:

		fMet	Arg	Ala	Phe	Ser
A-protein:	AUUCCUAGGAGGUUUGACCU.	AUG.	CGA.	GCU.	UUU.	AGU. G

		fMet	Ala	Ser	Asn	Phe
Coat-protein:	UAGAGCCCUCAACCGGGGUUUGAAGC.	AUG.	GCU.	UCU.	AAC.	UUU

		fMet	Ser	Lys	Thr	Thr	Lys
RNA replicase:	AAACAUGAGGAUUACCC.	AUG.	UCG.	AAG.	ACA.	ACA.	AAG

Fig. 4–2 Initiation sites with assigned amino acids for mRNA from R17 bacteriophage. Possible termination codons prior to the initiating AUG are underlined. ∿∿∿ shows a Shine and Dalgarno recognition site.

The currently generally accepted mechanism for binding the 30 S subunit arose from a proposal by Shine and Dalgarno from a determination of the nucleotide sequence of the 3′-end of the 16 S rRNA, the RNA component of the 30 S ribosomal subunit. Part of the sequence was noted to be complementary with a purine-rich region (e.g. AGGAGG for the A-protein) prior to the initiating AUG of determined initiation sites such as depicted in Fig. 4–2. This means that the mRNA initiation site is somehow located by the structural RNA of the ribosome. The Shine and Dalgarno proposal has been established through the correlation of information from numerous mRNA sequences whose pre-AUG sequences are known, and through experimental results, chiefly from Steitz's group.

The situation appears to be different in eukaryotes because many of these monocistronic mRNAs have no Shine and Dalgarno sequences and very short pre-AUG sequences. Possibly, in these cases, the 40 S subunit just binds to the 5′-beginning of the mRNA.

4.3   The termination signal

Polypeptide chain termination is defined as the mRNA-programmed release of free polypeptide from polypeptidyl-tRNA growing on the ribosomal surface. When homopolyribonucleotides such as poly U or poly A are used to programme the synthesis of a polypeptide in a cell-free system, nearly all of the polypeptide product can be isolated attached to tRNA. The small amount of chemical hydrolysis of the ester linkage in peptidyl-tRNA under physiological conditions is nowhere near adequate to explain the rate at which the bacterial cell completes proteins so there must be a mechanism for hydrolysing polypeptidyl-tRNA. It is now known that mRNA contains signals for polypeptide chain termination; the codons UAA, UAG and UGA have been shown to be the bacterial chain-terminating codons by genetic and biochemical methods. Although UAA and UAG are chain-terminating codons in eukaryotes, it is not clear, as observed previously, whether UGA is one.

Studies of the head protein fragments caused by amber mutants of T4 bacteriophage showed that the mutants lead to efficient termination of the polypeptide chain. This was suggestive evidence that nonsense codons code for proper chain termination. Furthermore, since the UAA codon was much more difficult to suppress than the UAG, UAA was considered to be the normal chain terminating codon whereas UAG could be a rare chain terminator. Genetic and biochemical studies have implicated another codon, UGA, as a terminator codon, also by characterization of a type of suppressible nonsense mutation. No bacterial aminoacyl-tRNAs which can be coded by UAA, UAG or by UGA have so far been found, although some eukaryotic aminoacyl-tRNAs are apparently able to decode the chain terminator codon UGA. Eukaryotic suppressor tRNAs for UAA and UAG have been isolated from mutant yeast cells.

The classical cases of suppression as in *E. coli* involve tRNAs with base changes in their anticodons so that they become able to decode a chain terminator codon. When a normal *E. coli* cell which is suppressor minus (su⁻) becomes changed into a suppressor positive cell (su⁺), as for example subclass three, i.e. $su_3^- \rightarrow su_3^+$, we know that there is a base change converting the G*UA (G* is a modified G) to CUA so that a su⁻ tRNA^Tyr decoding UAU/C changes to read the amber codon UAG (NB. anticodons are complementary and antiparallel with codons). Recently, other types of suppression have become apparent. One interesting type, not yet fully understood, may be a more efficient way of using genetic information. For example, cases are now known regarding bacteriophage genes where occasional readthroughs of normal chain termination codons at the ends of genes produce an extra population of elongated gene products, which can also have biological function. Indeed, in the case of Qβ bacteriophage, a readthrough product of the A or maturation protein produces a larger version of A called A1 which is a protein necessary for infectivity. Thus, the same gene can produce more than one gene product. It is rather doubtful if we can call this natural readthrough of termination actual suppression in the classical mutational sense, and we are not sure of the mechanism. Furthermore, more natural readthrough cases have been discovered in the translation of plant viral mRNA such as from tobacco mosaic virus, turnip yellow mosaic virus and also perhaps in the case of mammalian globin mRNAs. Currently, this type of research is attracting much interest in an attempt to see whether this really is another way of using genetic information more efficiently.

Whether natural chain termination *in vivo* involves a structural signal involving more than the terminator codons is not known; *in vitro* a terminator codon suffices. Very little is known about *in vivo* chain termination signals. Much of the evidence available so far comes from sequence studies on phage RNA; for example the nucleotide sequences at the ends of the coat-protein cistrons in two phage, R17 and f 2, are known. Both sequences (Fig. 4–3) show that codons for the last amino acid of the protein are followed not by just one terminator codon but by two in tandem (NICHOLS and ROBERTSON 1971). Recent work, however, makes it unlikely that this double codon sequence

End of coat-protein

Start of RNA
replicase cistron

Ala Asn Ser Gly Ile Tyr   fMet Ser

f₂.(G)CA AAC UCC GG⎡C⎤AUC UAC <u>UAA UAG A</u>⎡C⎤G CCG GCC AUU CAA ACA UG
R17.(G)CA AAC UCC GG⎣U⎦AUC UAC <u>UAA UAG A</u>⎣U⎦G CCG GCC AUU CAA ACA UGA GGA UUA CCC AUG UCG

Fig. 4–3 Intercistronic nucleotide sequence for bacteriophage R17 and f2 RNAs. Differences between the two phages are blocked. The termination codons are underlined. (Adapted from NICHOLS and ROBERTSON. 1971.)

UAAUAG is a universal termination signal. One explanation of this double terminator sequence is that it might protect against spontaneous mutations capable of converting a single termination codon into a sense codon. The double terminator codon makes it very unlikely indeed that a chance mutation could suppress termination.

A comparison of Fig. 4–2 with Fig. 4–3 reveals the unexplained property of some of the sequences known to precede the proper initiation codon AUG; in Fig. 4–2 there are possible terminator codons preceding the AUG but these are not at the cistron ends as can be seen from the determined location in Fig. 4–3.

4.4   Termination step of protein synthesis

Because mutant bacterial aminoacyl-tRNAs (suppressor tRNAs) which prevent protein chain termination *in vivo* and *in vitro* can be isolated and studied in cell-free systems, it was assumed that a special chain terminating tRNA decoded the termination signal. This idea delayed the discovery of the currently accepted idea of how chain termination takes place. In contrast to the mechanism of chain initiation it is now known that no tRNA is involved. However, complex protein factors called release factors are known to be involved in the hydrolysis of the peptidyl-tRNA triggered by a chain termination codon. Another difference from chain initiation is that the terminator codons do not specify other amino acids according to the position in the mRNA's sequence.

Elucidation of the detailed mechanism of peptidyl-tRNA hydrolysis during polypeptide chain termination is difficult because of the involvement of the ribosome whose structure will be of unresolved complexity for some time. Ribosomal binding of a release factor is not sufficient to trigger the hydrolysis of peptidyl-tRNA but there is reasonable evidence that one participates in the hydrolytic activity. It is not yet clear whether the hydrolysis is due to a co-operative interaction of release factors and a ribosomal constituent. It has, however, been shown that the ribosome must be in an active form (i.e. able to form peptide bonds) for the release activity to be possible.

A recent attractive theory is that chain termination is brought about by the peptidyl transferase enzyme causing hydrolysis rather than synthesis under the special influence of a termination signal and release factors. The cofactor GTP may also be involved in this step as well as in the other steps of protein biosynthesis. Whether this is true and whether water could be used as a substrate at the ribosomal A-site by the peptidyl transferase to hydrolyse the peptidyl-tRNA bound in the P-site are subjects of current research.

Progress towards extending our knowledge of the nature of the genetic code and control of protein biosynthesis clearly depend on gaining more information on the cellular architecture and macromolecular interactions. Our present methodology limits the possibility of acquiring new knowledge in this field but when the problem becomes clear, man's ingenuity usually comes to the rescue.

4.5   Conclusion

Various attempts have been made by theoreticians to find underlying rules in the codon assignments shown in Table 2–2. By observation it is seen that in general the left-hand side of the table contains the more nonpolar (hydrophobic) amino acids. The origin of the genetic code is related to the origin of life so has stimulated many attempts at trying to understand these basic biological problems. There is no clear, convincing answer to these problems. For the advanced reader the difficulties of theorizing on such a problem, when we see only a very efficient end product code which has excluded all others, can be pursued in the papers by CRICK, 1968, and ORGEL, 1968.

Efforts to prove the universality of the bacterial genetic code stimulated the search for eukaryotic cell-free systems active for synthesizing proteins. Results obtained from recently constructed systems such as those from rat liver, reticulocytes or ascites tumour cells, support the contention that the code is indeed universal for the various cellular cytoplasms. Furthermore, now that messenger RNA nucleotide sequences are being determined, it is possible to decide which codon of a degenerate set is used for a particular amino acid and how often. Different frequencies of codon usage may turn out to be significant in the possible regulation of protein synthesis by use of certain codons for aminoacyl-tRNAs which occur in controllable small amounts.

Bibliography

General reading – books and review articles

ADAMS, R. L. P., BURDON, R. H., CAMPBELL, A. M., LEADER, D. P. and SMELLIE, R. M. S. (1981). *The Biochemistry of the Nucleic Acids*, 9th edition. Chapman and Hall. A good basic background to the field including newer techniques of genetic manipulation.

BULL, A. T., LAGNADO, J. R., THOMAS, J. O., and TIPTON, K. F. (Eds) (1974). *Companion to Biochemistry*. Longman, London. Chapters 1, 2 and 7 for biochemical and chemical research evidence.

LEHNINGER, A. L. (1982). *Principles of Biochemistry*. Worth Publications, New York. An up-to-date readable textbook for general biochemistry and molecular genetics.

STRYER, L. (1981). *Biochemistry*, 2nd edition. W. H. Freeman & Co., San Francisco. An excellent general textbook covering up-to-date developments in the field with very pretty illustrations.

SZEKELY, M. (1980). *From DNA to Protein*. MacMillan Press Ltd, London; reprinted with corrections in 1981. A more detailed coverage with detailed bibliography of the subjects outlined in this text.

The Molecular Basis of Life. Readings from the Scientific American, W. A. Freeman, Reading. Very readable introduction to the field.

Annual Reviews of Biochemistry. Annual Reviews Inc., Palo Alto. A sound series for the active research worker to keep abreast of work in the field.

For experimental details about procedures one series can be especially recommended:

Methods in Enzymology. Academic Press, New York. Special volumes on nucleic acids and protein synthesis, eds. K. Moldave and L. Grossman, Vol. XIIA, Vol. XIIB, Vol. XX, Vol. XXI, Vol. XXIX, Vol. XXX, Vol. LIX, Vol. LX, Vol. LXV. A summary of these most useful techniques is found in *RNA and Protein Synthesis*, ed. K. Moldave, 1981, Academic Press.

Advanced Reading

Chapter 1

DREW, H. R. and DICKERSON, R. E. (1981). Structure of a B-DNA Dodecamer. *Journal of Molecular Biology*, **151**, 535–56.

WANG, A. H. -J., QUIGLEY, G. J., KOLPAK, F. J., CRAWFORD, J. L., van BOOM, J. H., van der MAREL, G. and RICH, A. (1979). Molecular structure of a left handed double helical DNA fragment at atomic resolution. *Nature, London*, **282**, 680–6.

Chapter 2

Cold Spring Harbor Symposium on Quantitative Biology (1966). The Genetic Code, Vol. XXXI. Contains many excellent papers by leading workers in the field.

CRICK, F. H. C. (1966). The genetic code III. *Scientific American*, **215**, Oct., 55–62.

HALL, B. D. (1979). Mitochondria spring surprises. *Nature, London*, **282**, 129–30.

Maverick mitochondria (1980). News & Views. *Nature, London*, **287**, 9–10.

SANGER, F., BARRELL, B. and colleagues (1977). Nucleotide sequence of bacteriophage φX174 DNA. *Nature, London*, **265**, 687–95.

STRETTON, A. O. W. (1965). The genetic code. *British Medical Bulletin*, **21**, 229–35. A good source of references for the early genetic work.

Chapter 3

ALTMAN, S. (Ed.) (1978). *Transfer RNA*. M.I.T. Press, Cambridge, Massachusetts.

CHAMBLISS, G., CRAVEN, G. R., DAVIES, J., DAVIS, K., KAHAN, L. and NOMURA, M. (Eds) (1980). *Ribsomes; Structure, Function and Genetics*. University Park Press, Baltimore. A necessary starting point for research workers to get up-to-date with a broad basis in the field of protein biosynthesis. Also appropriate for Chapter 4.

CLARK, B. F. C. (1980). The elongation step of protein biosynthesis. *Trends in Biochemical Sciences*, **5**, 207–10.

SCHIMMEL, P. R., SÖLL, D. and ABELSON, J. N. (Eds) (1979). *Transfer RNA: Structure, Properties and Recognition*. Cold Spring Harbor Monograph Series.

(1980) *Transfer RNA: Biological Aspects*. Cold Spring Harbor Monograph Series.

Two volumes giving the interested researcher a detailed foundation in the tRNA field and including new sequencing methods.

Chapter 4

CLARK, B. F. C. and MARCKER, K. A. (1968). How proteins start. *Scientific American*, **218**, Jan., 36–42.

CASKEY, C. T. (1980). Polypeptide chain termination. *Trends in Biochemical Sciences*, **5**, 234–7.

GAREN, A. (1968). Sense and nonsense in the genetic code. *Science*, **160**, 149–59.

GRUNBERG-MANAGO, M. and GROS, F. (1977). Initiation mechanisms of protein synthesis. *Progress in Nucleic Acid Research and Molecular Biology*, **20**, 209–84. Academic Press.

HUNT, T. (1980). The initiation of protein synthesis. *Trends in Biochemical Sciences*, **5**, 178–81.

References for the text

BARRELL, B. G. (1971). Fractionation and sequence analysis of radioactive nucleotides. *Procedures in Nucleic Acid Research*, Vol. 2, pp. 751–79 (G. L. Cantoni and D. R. Davies, eds). Harper and Row.

BARRELL, B. G. and CLARK, B. F. C. (1974). *Handbook of Nucleic Acid Sequences*. Joynson-Bruvvers, Oxford.

❋ CHAPEVILLE, F., LIPMANN, F., von EHRENSTEIN, G., WEISBLUM, B., RAY, W. J. and BENZER, S. (1962). On the role of soluble ribonucleic acid in coding for amino acids. *Proceedings of the National Academy of Sciences, U.S.A.*, **48**, 1086–92.

CLARK, B. F. C., DOCTOR, B. P., HOLMES, K. C., KLUG, A., MARCKER, K. A., MORRIS, S. J. and PARADIES, H. H. (1968). Crystallization of transfer RNA. *Nature*, **219**, 1222–4.

CLARK, B. F. C. and MARCKER, K. A. (1966). The role of N-formylmethionyl-sRNA in protein biosynthesis. *Journal of Molecular Biology*, **17**, 394–406.

CRICK, F. H. C. (1966). Codon–anti codon pairing: The wobble hypothesis. *Journal of Molecular Biology*, **19**, 548–55.

CRICK, F. H. C. (1968). The origin of the genetic code. *Journal of Molecular Biology*, **38**, 367–79.

CRICK, F. H. C. (1970). Central dogma of molecular biology. *Nature, London*, **227**, 561–3.

FRANKE, C., EDSTRÖM, J. E., McDOWALL, A. W. and MILLER, O. L. Jr. (1982). Electron microscopic visualization of a discrete class of giant translation units in salivary gland cells of *Chironomous tentans*. *EMBO Journal*, **1**, 59–62.

HOLLEY, R. W., APGAR, J., EVERETT, G. A., MADISON, J. T., MARQUISEE, M., PENSWICK, J. R. and ZAMIR, A. (1965). Structure of a ribonucleic acid. *Science*, **147**, 1462–5.

KIM, S. H., SUDDATH, F. L., QUIGLEY, G. J., McPHERSON, A., SUSSMAN, J. L., WANG, A. H. J., SEEMAN, N. C. and RICH, A. (1974). Three-dimensional tertiary structure of yeast phenylalanine transfer RNA. *Science*, **185**, 435–9.

MARCKER, K. A. and SANGER, F. (1964). N-formylmethionyl-sRNA. *Journal of Molecular Biology*, **8**, 835–40.

MILLER, O. L. (1973). The visualization of genes in action. *Scientific American*, **228**, Mar., 34–42.

NICHOLS, J. and ROBERTSON, H. (1971). Sequences of RNA fragments from the bacteriophage f2 coat protein cistron which differ from their R17 counterparts. *Biochimica Biophysica Acta*, **228**, 676–81.

NIRENBERG, M. W. and MATTHAEI, J. H. (1961). The dependence of cell-free protein synthesis in *E. coli* upon naturally occurring or synthetic polyribonucleotides. *Proceedings of the National Academy of Sciences, U.S.A.*, **47**, 1588–602.

NIRENBERG, M. W. and LEDER, P. (1964). RNA codewords and protein synthesis. *Science*, **145**, 1399–407.

ORGEL, L. E. (1968). Evolution of the genetic apparatus. *Journal of Molecular Biology*, **38**, 381–93.

RICH, A. and KIM, S. H. (1978). The three-dimensional structure of transfer RNA. *Scientific American*, **238**, Jan., 52–62.

ROBERTUS, J. D., LADNER, J. E., FINCH, J. T., RHODES, D., BROWN, R. S., CLARK, B. F. C. and KLUG, A. (1974). Structure of yeast phenylalanine tRNA at 3 Å resolution. *Nature, London*, **250**, 546–51.

SPRINZL, M. and GAUSS, D. H. (1982). Compilation of tRNA sequences. *Nucleic Acids Research*, **10**, r1–r55.

STEITZ, J. A. (1969). Polypeptide chain initiation: nucleotide sequences of the three ribosomal binding sites in bacteriophage R17 RNA. *Nature, London,* **224**, 957–64.

TERZAGHI, E., OKADA, Y., STREISINGER, G., EMRICH, J., INOUYE, M. and TSUGITA, A. (1966). Change in a sequence of amino acids in phage T4 lysozyme by acridine-induced mutations. *Proceedings of the National Academy of Sciences, U.S.A.,* **56**, 500–7.

Index